A NEW AGENDA FOR SUSTAINABILITY

Ashgate Studies in Environmental Policy and Practice

Series Editor: Adrian McDonald, University of Leeds, UK

Based on the Avebury Studies in Green Research series, this wide-ranging series still covers all aspects of research into environmental change and development. It will now focus primarily on environmental policy, management and implications (such as effects on agriculture, lifestyle, health etc), and includes both innovative theoretical research and international practical case studies.

Also in the series

At the Margins of Planning
Offshore Wind Farms in the United Kingdom
Stephen A. Jay
ISBN 978 0 7546 7196 1

Contentious Geographies
Environmental Knowledge, Meaning, Scale
Edited by Michael K. Goodman, Maxwell T. Boykoff and Kyle T. Evered
ISBN 978 0 7546 4971 7

Environment and Society
Sustainability, Policy and the Citizen
Stewart Barr
ISBN 978 0 7546 4343 2

Multi-Stakeholder Platforms for Integrated Water Management
Edited by Jeroen Warner
ISBN 978 0 7546 7065 0

Protected Areas and Regional Development in Europe
Towards a New Model for the 21st Century
Edited by Ingo Mose
ISBN 978 0 7546 4801 7

Energy and Culture
Perspectives on the Power to Work
Edited by Brendan Dooley
ISBN 978 0 7546 4514 6

A New Agenda for Sustainability

Edited by

KURT AAGAARD NIELSEN
Roskilde University, Denmark

BO ELLING
Roskilde University, Denmark

MARIA FIGUEROA
Technical University of Denmark, Denmark

and

ERLING JELSØE
Roskilde University, Denmark

Routledge
Taylor & Francis Group

LONDON AND NEW YORK

First published 2010 by Ashgate Publishing

Published 2016 by Routledge
2 Park Square, Milton Park, Abingdon, Oxfordshire OX14 4RN
711 Third Avenue, New York, NY 10017, USA

First issued in paperback 2016

Routledge is an imprint of the Taylor & Francis Group, an informa business

British Library Cataloguing in Publication Data
A new agenda for sustainability. -- (Ashgate studies in
   environmental policy and practice)
   1. Sustainable development. 2. Environmental policy.
   I. Series II. Nielsen, Kurt Aagaard, 1948-
   363.7'0561-dc22

Library of Congress Cataloging-in-Publication Data
A new agenda for sustainability / by Kurt Aagaard Nielsen...[et al.].
   p. cm. -- (Ashgate studies in environmental policy and practice)
   Includes index.
   ISBN 978-0-7546-7976-9 (hbk)
1. Sustainable development. 2. Sustainable living. 3. Human ecology. 4. Environmental policy. I. Nielsen, Kurt Aagaard, 1948-
   HC79.E5N4495 2010
   338.9'27--dc22

                                                                              2010025242

ISBN 13: 978-1-138-24809-0 (pbk)
ISBN 13: 978-0-7546-7976-9 (hbk)

Contents

List of Figures

List of Tables

Preface

This book arose out of an international conference on a new agenda for the sustainable development of contemporary societies. The conference was held at Roskilde University in December 2007 and specific themes for the discussions at the conference were all based on papers presented by different faculty members of the Department of Environmental, Social and Spatial Change, Roskilde University. It is these papers – edited and revised on the bases of the discussions at the conference and reviews from outside scholars and experts – that appear in the book. Since the Department is interdisciplinary and deals with nearly all aspects of reality that can be connected to sustainability: natural, social, political, economic and environmental, the following chapters focus on a wide range of issues and subjects that are considered in common to constitute new ways and new agendas for sustainable development. Thus the title of the book *A New Agenda for Sustainability* should not be understood as if we attempt to show *the* way or *the* agenda. We wish to present a diversity of subjects and approaches that all reflect on the necessity for change. This requires a better understanding of a complex reality as well as maintaining and strengthening the hopes and aspirations for a better future in environmental, social and economic terms.

The book has been long underway and we acknowledge efforts, facilitation and help from many persons. All articles were reviewed by an international highly qualified panel of scientists. We are highly grateful to the following reviewers: Jan Tjeerd Boon, Trygver Bjørknes, Christian Coff, Andrew J. Dugmore, Lars Emmelin, Jukka Gronow, Henrik Gudmundsson, Rolf Guttesen, Echardt Hildebrandt, Marco Joas, Robert Karasek, Søren Kerndrup, Mari Kira, Ortrud Lessman, Britha Mikkelsen, Torben Hviid Nielsen, Nadarajah Sriskandarajah, Joel Tichner, Edvardo Vasconcellos and Thomas Whiston.

Andrew Crabtree has not only contributed to the book as an author, but also carried out English reviews of all the articles. Ritta Bitsch, technical assistant at the Department, has made an excellent job of preparing the manuscript for printing and we are also highly grateful to the publishing staff at Ashgate. Finally, we would like to thank all the authors for their ongoing enthusiasm, perseverance and patience on the long road.

The Editors
October 2009

Introduction
A New Agenda for Sustainability

Maria Figueroa, Bo Elling, Erling Jelsøe and Kurt Aagaard Nielsen,

The last three decades have seen intense and wide-ranging debates concerning the concept of sustainable development, its usage and operationalisation. These debates have taken place both in international scholarly and public policy arenas, and they remind one of the evolutions of other concepts such as democracy, freedom or welfare. Disagreements exist even among those who support and encourage the use of the term and result from differences in theoretical approaches, its interpretation at policy level or in its philosophical and ethical implications.

At the same time, these differences have presented opportunities, propagated analytical models, value discussions and extensive research on the conditions for achieving sustainable development. The present book wishes to acknowledge this pluralistic tradition while addressing concrete empirical approaches to the debate on sustainability.

The major motivations for the book are the need for the renewal of the concept and for the initiation of new practices, two decades after the Brundtland Commission's Report, *Our Common Future* (World Commission on Environment and Development 1987) adopted the most widely used definition of the concept as 'development that meets the needs of the present without compromising the ability of future generations to meet their own needs' (World Commission on Environment and Development 1987: 8).

Our purpose is twofold. First, to provide the so-called 'future generation perspective', which was a central but unresolved notion in the original Brundtland definition and, second, to provide a review of the new milestones for policy and research that can expand the discussion and point to future conceptualisations of sustainability.

Structure and Application of the Book

The differences in the contributions reflect the importance of interdisciplinary and transdisciplinary work in the field and, more strategically, they attempt to provide new parameters and perspectives for discussions that can inspire future work on global sustainability, be it political or academic.

To organise the highly diverse contributions in the volume, we have clustered chapters in relation to the topics they address. We have avoided a policy-oriented context even though several chapters deal with issues within sectors such as health, agriculture and transportation in which the problem of sustainability is particularly urgent. Our choice has been to assemble related chapters in a way that facilitates discussion of their potential convergences and interconnections. In doing so, the book has been divided into three parts.

The first part is concerned with philosophical, ethical and meta-theoretical aspects of the concept of sustainability. They delineate the conditions for strengthening and expanding the concept of sustainability, and envision how long-lasting, democratically driven societal changes aimed at attaining sustainable development can be initiated and supported.

The second part approaches sustainable development as a strategic field focusing on what sustainability could and should be in reference to institutional and administrative competences. Furthermore, the concept is analysed in relation to planning processes that could lead towards a transition to a sustainable society in general and towards the sustainability of the production and consumption systems in particular.

In the third part, sustainability is viewed from 'everyday life' and 'life politics' perspectives emphasising the complex and important relationship between democracy and sustainability. While individual citizens have gained growing awareness of sustainability in their practices, actions and cultural undertakings, scaling up these efforts to achieve a sustainability transition will require a strong, well-founded, focused and continued democratic dialogue, supported by strategic alliances and networking. Such developments require new types of political modernisation and forms of transitional governance.

All the chapters in the book aim to achieve these purposes and ambitions and as such they are part of a common body of research. In our editorial work, we have tried to facilitate this through a continuous dialogue with and amongst the authors; though the opinions expressed are the authors and not the editors.

The range of research activities supporting this volume encompasses a broad range of disciplines and approaches from the natural and environmental sciences, and the social and the political sciences and the humanities. In this effort we have strived to understand modern society's opportunities for long-term cohesion and the key role that cultural and social aspects play in society's long-term utilisation of resources and environmental management. Thus, we hope that the book can become a valuable reference for a diversity of scholars working on sustainable development.

The authors of the work presented here belong to different international networks and so the book as a whole has a unique possibility to open a conceptual dialogue concerning third generation sustainability studies. The book is written for scholars and researchers who find the conceptual dialogues about sustainability important for the future of the globe.

Relation to Existing Literature

According to the Brundtland Commission, the notion of sustainable development contains two key concepts or ideas:

the concept of 'needs', in particular the essential needs of the world's poor, to which overriding priority should be given; and the idea of limitations imposed by the state of technology and social organisation on the environment's ability to meet present and future needs. (World Commission On Environment And Development 1987: 43)

This approach to sustainability puts emphasis on at least three important dimensions that have been discussed extensively in the literature (Redcliff 1987, Pezzey 1992, Lafferty 2004). First, it places human beings and human welfare above environmental or ecological sustainability. Second, it introduces an element of intra- and inter-generational justice: global equity among living generations and of generational equity with respect to future generations. Third, it offers constraint dimensions to which human development must be assessed with respect to the limits of nature and overall global ecological balance, so-called 'physical sustainability' (Lafferty 2002: 266). The general position in the book is that these dimensions provide a core set of values, norms and goals capable of creating standards for political action and change that make the Brundtland concept of sustainable development the best available or, put it another way, worthy of re-endorsement. Different opinions regarding the agencies to which these dimensions can be applied still underlie most of the controversies around the utilisation of the concept. For example, which dimensions are relevant at company level, at the level of political decision making and at the level of life politics.

Sustainable development as a concept will continue to be contested and marked by contradictions. As Redclift pointed out and explored as early as 1987, almost before Brundtland's World Commission on Environment and Development published its report (Redclift 1987), the contradictions primarily relate to the concept of 'development' and its implications for the environment. Similarly, Langhelle (2002) has also stated that the coupling of development and the environment means that two ideologically contested concepts have been added together to create a third. Finally, Nielsen and Nielsen have stressed that despite the fact that 'sustainable development' sounds quite innocent; it nevertheless implies a de-radicalisation of the concept of sustainability because 'development' is part of the logic of modernisation according to which development simply means the propagation of a capitalist market economy. In this way, the meaning of sustainable development has become ambiguous (Nielsen and Nielsen 2006: 265).

At the turn of the new millennium, discussions about the notion of sustainability shifted to assign greater importance to the role of economics and the links between environmental and economic values in guiding societal decision making. At the heart of this predicament has been the idea that the environment is misused and degraded because its full value does not get reflected in economic markets (Norgaard 1992). From this perspective, sustainability criteria can be applied to any system (project, political decision, industrial) as long as an appropriate economic valuation of the environmental goods is established and incorporated as part of a cost-benefit appraisal. This process held the promise of improving the way in which economies and markets take care of environmental problems.

The prominence that cost-benefit thinking attained, despite numerous early critiques,[1] came to substitute the Brundtland concept. Now sustainability is not seen as a notion providing a new *qualitative* understanding of growth. Cost-benefit brought discussions back to quantitative measures of growth.

As a result of the emphasis and direction provided by the cost-benefit approach, attempts were made to improve the role of economic decision making in relation to sustainability. One of those strategies was ecological modernisation (Weale 1992, Hajer 1995). In John Dryzek's words, ecological modernisation put forward the idea that a clean environment is good for business and that such an agenda would have to be advanced as a cultural push within institutions, political constituencies and consumers (Dryzek 2004).

In an analogous effort, the ecological aspects of the sustainable development debate were emphasised through the introduction of concepts such as 'eco-development' advanced by Wolfgang Sachs of the German-based Wuppertal Institute. The notion of eco-development attempted to supplement the formulation of sustainability criteria while giving priority to 'social and ecological sound management fostered in a spirit of solidarity that places importance on new forms of lifestyle linked to minimal material production and consumption' (Sachs 1988).

An even more radical, indigenous approach to ecological and social sustainability has been forwarded by Vandana Shiva from the Indian Institute of Management in Bangalore. She is critical of the idea of development as an attempt to impose alien Western imperatives on the Third World and conceives sustainable development as part of an effort to achieve global democratic utilisation of natural resources (Shiva 2006).

In different ways these messages highlight the contradictions and lack of clarity concerning the concept of sustainability. However, the different emphasis and approaches to the concept among economists, ecologists and environmental ethicists soon led to a sharp separation between those favouring a pro-growth or 'sustainable growth' stance and those who have a more critical approach to growth seeking to prioritise environmental sustainability. Remarkably, throughout the controversy, the terms of the debate have remained under the general 'framework' created in the Brundtland meta-objective of sustainable development.

---

1   As early as 1985, strong criticisms had already been put forward showing the weaknesses of the cost-benefit approach when dealing with environmental problems. For instance, Norgaard (1985) argued that some of the assumptions underlying cost-benefits analysis goods could be problematic or unrealistic such as: 'the assumption that environmental systems are divisible and can be owned; that environmental systems can achieve an equilibrium position; and that changes to systems are reversible'. Norgaard also noted concrete examples of how it may be hard to account for multiple pollutants where chemical and biological changes are such that the combined effect is very different from the sum of two or more individual components (Ibid.).

The effect on policy making of the different emphases has been quite significant. In particular, the neo-liberal trend that has gained momentum since the beginning of the new millennium successfully substituted the terms of the sustainability debate within policy making from the qualitative notion found in the Brundtland Report into a more quantitative approach that promised the possibility of obtaining greater sustainability for any given amount of investment. This notion did not seek to challenge economic growth but saw it as the unequivocal parameter within which sustainable development ought to be framed. This trend gained acceptance amongst neo-liberal political circles around the world and substantially contributed to the renewal of economic thinking in Denmark where Bjorn Lomborg, author of *The Sceptical Environmentalist* (Lomborg 2001) has been an influential political figure representing this direction.

This economic thinking was also represented by authors such as Paul Hawken and Amory Lovins in the US, who developed concrete ideas for radical changes toward an economically efficient ecological modernisation with the publication of *Natural Capitalism* (Hawken et al. 1999). They argue that technological modernisation can lead to a new strategy of capital accumulation for private firms and efficient growth that they term 'green capitalism'. In this ecological modernisation process, the qualitative notion of sustainability taken from the Brundtland Report disappeared, leaving in its place a form of deregulated concept of sustainability which, in our view, detached societal actors from moral considerations and substantial direct responsibility.

In the Brundtland Report, sustainability included qualitative aspects of development such as livelihoods of the poor at the same level as responsibility for natural resources. The qualitative aspects of sustainability gave room for the later prioritisation of the notion of 'corporate social responsibility' (CSR), introduced by the Business Council for Sustainable Development (1992). This defined the role of business in sustainability matters within a general strategy of capital accumulation. The concept of social responsibility was touched upon in the Brundtland Report, though it played a less prominent role in the overall discussion on sustainability.

The new business-oriented agenda strengthened polarised understandings of the concept into strong and weak forms of sustainability. Strong sustainability insisted on seeing nature as an absolute need (authors include Meadows et al. 1972, Lester Brown of the World Watch Institute, William Rees, who developed the concept of 'Ecological footprint', Wolfgang Sachs, Vandana Shiva and deep ecologists such as Norwegian philosopher Arne Næss), while weak sustainability advocates insisted on the substitution processes in which scarce resources could be replaced by new, and perhaps not yet known, resources.

The proposers of weak sustainability have confidence in the potential that innovation can bring to society through basic market mechanisms. The implication is that a perfect interaction between technologies and well-functioning markets will make apparent the changes that are required if society needs to move in the direction of sustainability. However, this idea is currently being challenged for producing too few results too late. Particularly as mounting evidence has been gathered on the linkages between global climate disruption and human activity. It has become clear that relying solely on signals from the interaction between markets and technology to provide the needed changes in a sustainable direction will not be sufficient.

Today, a trend can be seen indicating a third phase of sustainability thinking. This trend has become more apparent following a number of serious and devastating events which are potentially linked to climate change-related hazards. In our view, this trend gained significant momentum with the selection of Al Gore and the Inter Governmental Panel for Climate Change (IPCC) as joint winners of the Nobel Peace Prize, and the publication of the latest IPCC report in January 2007, and the widely publicised Stern Review Report (Stern 2006). These events have played an important role in helping define the case for urgent action to stabilise CO_2 emissions and to emphasise the specific societal responsibility that these actions call for. However, it is too early to describe the way in which this trend and the larger discussion about climate change will ultimately embrace the concept of sustainability.

The emerging sides of the current debate demonstrate that the validity of a qualitative message for sustainability has endured and needs to be reconsidered. Different chapters presented in this book attempt to demonstrate the inextricable link between sustainable development and societal responsibility on the local, regional, national and global levels. The message we wish to put forward is that a new epoch – in fact a third generation of sustainability concepts – has begun, but not yet accomplished, the provision of new forms of holistic thinking on sustainable development.

Financial disruption was not the reality when most of the chapters were written, but somehow most of the chapters implicitly anticipate what has become even clearer today: we need more society and less reliance or trust in market mechanisms. Sustainability is about creating a society beyond markets and market relations.

Trails for a Common Perspective in the Book

A common thematic exists between all the contributions in this book that seeks to explicate how the concept of sustainability can be used and what problems can be associated with the existing definition. The authors have attempted to avoid a strong instrumental use of the concept, favouring a notion of sustainable development as an open and unfinished concept, just in the same manner as the notion of democracy is used to avoid all forms of fundamentalisms that tend to rule out complementary or opposing views. The openness of the concept of sustainability makes it a meeting point for many different actors in a common attempt to civilise the global society. Different actors will attach different meanings to the concept and have different views of the social processes leading towards sustainability.

A similar point was formulated by Douglas Torgerson (1994) in his critical discussion of the 'administrative mind' and the ideology of industrialisation as a basis and framework for a strategy leading to sustainability, when he argued for a decentred approach to sustainability and noted, with reference to Kai N. Lee (1993), that even though the idea of sustainability involves a comprehensive vision of the planet the process unfolds in a context of diversity. And in relation to the conceptual discussion about sustainability he concluded that the many meanings and connotations of sustainable development do not signal a 'neat concept', but a range of ideas whose very indeterminacy and contestability help to suggest previously unheard possibilities.

The pluralistic stance of the contributors in this book supplements a critical view of sustainability; critical in the sense that all the authors maintain a normative position toward the context of social problems of sustainability, wishing to show that sustainability cannot only be taken as a kind of sectorial politics, but it needs to be expanded within a basic daily societal order. Many different accounts are possible for this implicit daily societal order, but none of the articles addresses sustainable development as a neutral theme to provide a 'technology' or 'receipt' formula that can be acknowledged from different value positions. Sustainable development carries an unavoidable and strong political normative orientation linked to its field. This normative orientation we have attempted to explicate within the fields each of the authors have developed in their respective contributions. All attempt to emphasise an understanding of sustainability as a multifaceted and differentiated term.

There are no such fundamental categories as Mother Nature, from which science can derive interpretations of sustainability. Instead all chapters argue for a form of democratic sustainability which should allow opening up public debate around different visions and alternative future ways of organising our society politically, socially and culturally.

The chapters in the book are organised in three different levels or parts. In the following we introduce the levels of conceptualisation and the contributions of the entire book.

Part I: Challenging the Concept of Sustainable Development

In Part I, the conceptual aspects of sustainability are reflected on. The concept is challenged from a holistic position, though the authors are aware that holism can lead to fundamentalism unless the dynamics at stake in *the whole* are validated and connected to democratic culture. Different options for dynamics are presented and the notion of sustainability as a necessarily global issue is developed. Finally, the concept of sustainability is discussed in relation to societal rationalities/ethics.

Altogether the chapters in Part I express the idea that no one adequate, holistic concept of sustainability exists, however there are a number of concepts which can be seen as complementary and they are united in arguing that the issue of sustainability plays a central role in contemporary society.

According to Hvid, sustainability requires the understanding that life can be performed not only with thoughtfulness and reflection but also within practical forms: rhythms. The idea is that sustainability is not only concerned with environmental standards and issues of social justice which have received attention in democratic discussions concerning sustainability. It is also concerned with the practical rhythms in everyday life as embedded in social, economic and natural *activities*. Sustainability is about establishing sound rhythms and not just about reflexivity in our actions.

According to Elling, the concept of sustainability not only expresses the need of harmony between development and the environment, but the concept calls for the understanding of more profound, fundamental or structural contradictions in rationalities and discourses in society. He argues that sustainability must be connected to a concept of democratic cultural rationalisation. Elling urges us to realise that since modern society has no centre, planning for a sustainable future cannot be something controlled by societal authority. Sustainable development is necessarily a negotiated matter – something that needs to be agreed upon and requires reaching mutual understanding in public spaces.

Brandt discusses a nearly thousand-year-old carrying capacity system for sheep farming on the Faeroe Islands called 'skipan'. This traditional and nature-based ecological agricultural system is now challenged by modern farming methods which were introduced during the twentieth century along with a new socially-based sustainable fishing economy which has gradually developed during the century. Over the last few decades, this economy has been additionally supported by fish farming in the island fjords. This has also been based on social and short-sighted economic benefits which have not taken issues pertaining to long-term carrying capacity and norms into consideration. Against this historical background, Brandt highlights the contradictions between ecological and social sustainability, and explains how a social framing of sustainable development can benefit these different system-based concepts of sustainability.

Agger and Jelsøe reflect on the role of the ethical aspects of sustainable development. They discuss the way the Danish Council of Ethics, of which Agger is Chairman, dealt with a question about the use of the concept of utility as a criterion in the assessment of the deliberate release of genetically modified organisms (GMOs) into the environment.

The ethical aspects have, according to Agger and Jelsøe, often been overlooked. In connection with the discussion about sustainable development and GMOs, the notion of biodiversity is introduced as an important, related, concept. According to Agger and Jelsøe, ethical judgements can mediate between criteria from a horizon of utility and criteria from a horizon of biodiversity.

Part II: Sustainability in Relation to Political and Institutional Actions and Activities

In this part of the anthology, we focus on the challenges relating to political and institutional actions and activities if we are to achieve a more sustainable society. Problems of applying traditional managerial means in ensuring sustainable development reveal the need for new perspectives in the political and administrative processes. Government programmes and legislation at national and supranational (e.g. the EU) levels use sustainability as label for a wide range of changes of which some are at best incremental. This inflationary use of the concept, due to its positive connotations, raises questions about its meaning and significance in relation to more long-ranged and fundamental claims to change. The various contributions in this part of the anthology demonstrate that the problems of organising and planning for a more sustainable society are complex in nature, and involve barriers in many parts of the political and administrative processes. All in all, the idea of sustainability is a major challenge to the current political and administrative organisation of changes in society in general and in production systems in particular.

Hansen, Søndergård and Stærdahl reflect on the lack of success of the ecological modernisation of production systems. They suggest a different approach to the planning process taking their point of departure in a conception of changes at a socio-technical systems level. By emphasising the socio-technical systems level, and by acknowledging multiple 'generators' of change, the planning process becomes a matter of creating opportunities amongst different technologies and cultures in a system, rather than the top-down management of the production system. Instead of focusing on ends and management, this approach focuses on transitions – on changes in the system – and the possibilities for technology and culture to become more sustainable. Introducing the concept of transition arenas they discuss some of the problems and conflicts associated with governance of the transition towards sustainability.

By examining the background to the REACH (Registration, Evaluation, Authorisation and Restriction of CHemical substances) Agreement, Rank, Syberg and Carlsen demonstrate the challenges of promoting sustainability at an international, political, level – the EU. This is in part hampered by the additional complexities that arise from the political cross-cutting nature of sustainability. The formation of one of the most comprehensive EU policies demonstrates the various battles in a political process between different political and commercial interests. It starts with the agenda setting process of defining the general 'problems' of the chemicals in nature and continues to the decision-making process of defining sustainability in terms of specific damage to nature as risky consequences for production for the thousands of different chemicals.

Jelsøe and Kjærgård shed light on another critical problem in relation to sustainable development. The issue of sustainable agriculture and food production has been subject to extensive attention and debate. The chapter deals with the ambiguities and dilemmas in conceptualising sustainable agriculture and the authors examine some of the institutionalised conceptualisations of sustainable agriculture. The restrictive nature of the conceptions of sustainability is illustrated through a critical discussion and analysis of the Danish national strategy for sustainable development. The concept of food democracy is introduced as a way to involve citizens more directly in a development towards new and more sustainable practices, and establish pressures for a new form of social regulation within the food sector.

Crabtree examines the notion of development employed by the Brundtland Commission and argues that the concept of basic needs is fundamentally flawed and should be replaced by one of capabilities. Development is thus understood here as increasing the real freedoms that individuals have to lead the lives they value. After outlining the essential capabilities, both individual and collective, which are necessary for the leading of valued lives, including agency, the chapter continues by examining the issue of mental health, which restricts agency, a topic which is almost entirely absent from the sustainability debate. After documenting the importance of mental health, the chapter ends by providing a model for analysing the relationships between mental health and the environment.

Hansen addresses an issue which is quite different from those of the other chapters in this section, namely the difficulties of economic analysis in relation to sustainability. This is, however, an issue with important political and institutional implications due to the central role attributed to economic analysis in environmental policy. Basically, the problem has to do with the assumption that the perfect, frictionless, substitution of environmental values by economic values is possible, which is an assumption frequently used, e.g. in cost-benefit analyses. The chapter reviews the economic standard model for balancing environmental and economic objectives, and compares the assumptions behind the model with standard assumptions made in other academic disciplines.

Part III: Sustainability in an Everyday Life Perspective

The contributions in Part III explore paths for social change toward sustainability arising from more direct participatory forms of democracy in areas such as the management of national parks, urban environmental management, local transport and urban planning for sustainability. Two of the chapters outline challenges to policy and decision making resulting from alternative bottom-up urban civil society practices where people combine lifestyle choices and very focused and sometimes discontinuous participation and political activity, a style of public involvement that has been termed 'life politics' (Giddens 1991) or 'sub-politics' (Beck 1998). Participation for sustainability requires a public voice capable of articulating complex and often contradictory interests pertaining to matters affecting their communities and their immediate vicinity. Enlarging the franchise of participation does not guarantee progression towards sustainability goals since environmental interests do not always receive strong representation, while other citizen interests groups may gain a stronger position for advancing their own agendas.

The chapter by Holm reflects on numerous and varied experiences that participation in LA21 has achieved over the last 20 years after Rio, where the strongest and most successful experiences of strategic alliances and networking have remained functional even after the withdrawal of resources and attention. The path to achieving greater sustainability at a societal level requires 'scaling up' this 'niche' experience into a larger orchestrated effort. The chapter discusses how small signs of this have been taking place in Scandinavia, for example in organic food supply, purchasing systems, renewable energy systems, and waste and recycling infrastructure.

The chapter by Clausen, Hansen and Tind explores the challenges and contradictions resulting from a process in which the government actively sought to extend and improve public participation as a way of advancing sustainability goals in the management of national parks in Denmark. The resulting demands and visions that emerged from the process challenged conventional policy approaches for nature conservation held by the authorities and by the long-standing nature conservation organisation. However, the process failed to advance the needs of the citizens and their visions for the future.

Bransholm Pedersen and Land examine how health problems can become a common social responsibility. Health is a vital component of sustainability that needs to be brought to the forefront of debates and policy making. The concepts of health and sustainability can stimulate each other as a duality that can indicate the direction, the content and the quality of development that is to be sustained. They argue for a form of development that can secure better and equal conditions of health within sustainable patterns of production and consumption.

Figueroa suggests that the process of planning and experimenting to achieve sustainable transport systems in the Global South needs to consider accepting the informality, spatial chaos and intensely unruly nature of urban life as quasi permanent features that cannot simply be planned away. They affect the breadth and conditions upon which sustainable transport solutions should be approached. A change of expectation in research and planning is suggested, from one oriented towards achieving order and neatness, based on heavy investment and technologically-driven solutions; to one seeking to democratise mobility chartering a more plausible path to attain sustainable mobility to the majority of the urban poor.

References

Beck, U. 1997. *The Reinvention of Politics.* Cambridge: Polity Press.
Beck, U. 1998. *Democracy Without Enemies*. Cambridge: Polity Press.
Business Council for Sustainable Development. 1992. *Changing Course: A Global Perspective on Development and the Environment.* Cambridge, MA: Massachusetts Institute of Technology Press.
Dryzek, J. 1997. *The Politics of the Earth.* Oxford: Oxford University Press.
Dryzek, J. and Schlosberg, D. 2004. *Debating the Earth, The Environmental Politics Reader.* Oxford: Oxford University Press.

Hajer, M. 1995. *The Politics of Environmental Discourse: Ecological Modernization and the Policy Process*. Oxford: Clarendon Press.

Hajer, M. and Wagenaar, H. 2003. *Deliberative Policy Analysis: Understanding Governance in the Network Society*. Cambridge: Cambridge University Press.

Hawken, P., Lovins, A. and Lovins, L.H. 1999. *Natural Capitalism: Creating the Next Industrial Revolution*. Boston: Back Bay Books.

Lang, T. and Heasman, M. 2004. *Food Wars: The Global Battle for Mouths, Minds and Markets*. London/Stirling: Earthscan.

Langhelle, O. 2002. Bærekraftig utvikling, in *Samfunnsperspektiver på miljø og utvikling*, edited by T.A. Benjaminsen and H. Svarstad. Oslo: Universitetsforlaget, 225–54.

Lee, K.N. 1993. *Compass and Gyroscope: Integrating Science and Politics for the Environment*. Washington, DC: Island Press.

Lomborg, B. 2001. *The Sceptical Environmentalist: Measuring the State of the World*. Cambridge: Cambridge University Press.

Meadows, D.H. Meadows, D.L., Randers, J. and Behrens III, W.W.1972. *The Limits to Growth*. New York: Universe Books.

Nielsen, B.S. and Nielsen K.A. 2006. *En Menneskelig Natur. Aktionsforskning for Bæredygtighed og Politisk Kultur*. Copenhagen: Frydenlund.

Norgaard, R.B. 1985. Environmental Economics: An Evolutional Critique and a Plea for Pluralism. *Journal of Environmental Economics and Management*, 12(4), 382–94.

Pezzey, J. 1992. Sustainability: An Interdisciplinary Guide. *Environmental Values*, 1(4), 321–62.

Pretty, J. 1995. *Regenerating Agriculture: Policies and Practice for Sustainability and Self-reliance*. London: Earthscan.

Redclift, M. 1987. *Sustainable Development: Exploring the Contradictions*. London and New York: Routledge.

Sachs, W. 2000. *Development – the Rise and Decline of an Ideal*. Wupperthal Papers No. 108. Wupperthal: Wupperthal Institute for Climate, Environment and Energy.

Shiva, V. 2006. *Earth Democracy, Justice, Sustainability and Peace*. London: Zed Publications.

Stern, N. 2006. *The Stern Review Report on the Economics of Climate Change*. Cambridge: Cambridge University Press.

Torgerson, D. 1994. Strategy and Ideology in Environmentalism: A Decentered Approach to Sustainability. *Industrial and Environmental Crisis Quarterly*, 8(4), 295–321.

Weale, A. 1992. *The New Politics of Pollution*. Manchester: Manchester University Press.

Wilkinson, J. 2006. Fish: A Global Value Chain Driven onto the Rocks. *Sociologia Ruralis*, 46(2), 139–53.

World Commission on Environment and Development. 1987. *Our Common Future*. Oxford: Oxford University Press.

PART I

Challenging the Concept of Sustainable Design

Chapter 1
Sustainable Rhythms: When Society Meets Nature

Helge Hvid

Sustainability, the Grand Project for the Twenty-first Century

Sustainability is civilisation's main project for the twenty-first century, just as democracy was the new project of civilisation in the nineteenth century, and the welfare state was the new project of civilisation in the twentieth century in Europe and North America (Hvid 2006). It is still quite unclear what sustainability really is (as is also the case with 'democracy' and 'welfare'), however, numerous articles, reports and official documents state that sustainability is based on three pillars: environmental protection and regeneration, social protection and development, and economic regeneration and development.

The concept of sustainability is used in many contexts: international organisations, governmental institutions, non-governmental organisations (NGOs), local authorities, trade unions and, not least, in business. It has received many interpretations. However, the openness of the concept makes sustainability a possible meeting point for many different actors in a common attempt to civilise global society and create a (more) sustainable balance between human production and consumption, on the one hand, and regeneration of resources, both natural and human, on the other. We cannot expect that these different actors will work together in complete harmony for sustainability. On the contrary, there are a lot of conflicts related to sustainability. However, the fact that the actors are meeting creates a dynamic for sustainability of some kind.

Sustainability calls for big changes in the ruling systems: governmental rules must be changed, the priorities of business must be changed, the planning systems must be changed, new economic incentives must be developed, common values and cultures must be changed – a huge and perhaps unrealistic project? Optimism can be gained from the fact that changes in the ruling systems and in business are happening all the time. What the notion of sustainability is asking for is a new direction for these changes, namely towards sustainability. In a non-totalitarian society, such a shift in the patterns of social change can, however, not be realised without a broad acceptance by the population, and even more importantly, it cannot be realised without active support from daily life, in workplaces, in families and in local communities. Therefore it is crucial to create an understanding about what a sustainable everyday life is like. Where do we find the roots for sustainable rhythms in everyday life? What can make a sustainable everyday life attractive?

This chapter will not give a definitive answer to these big questions. It will, however, argue that the concept of rhythm is fruitful for finding the big answers. In this chapter I argue that the character of the rhythms of everyday life is crucial for sustainability. Rhythms are here understood as repeating activities in time and space – activities related to production, to consumption, to family activities, to local community, to cultural traditions, etc. Here I will define what sustainable rhythms are. In the rest of the chapter I will argue for this definition:

1. Sustainable rhythms are a combination of repetitive structure in time and space AND individual opportunities (for man, animals and plants) to differentiate, to adjust the rhythm, to adapt activities and to recreate the rhythm in new forms. That creates opportunities for regeneration of resources, survival, learning and development. The combination of repetition and discontinuity strengthen each living rhythmic circle.
2. Sustainable rhythms are interconnected with many other rhythms. That is what bio-diversity is about. That is what social diversity is about. That is what collaboration between different competences is about. That is what producer-customer relations are about. That is what a living local community is about. The interconnections between rhythms are sustainable when the one living circle creates resources for the other living circle: when the one species creates resources for another. When the one qualification creates resources for the other.

The claim is that even smaller changes in the rhythms of daily life could create bigger changes in the governing system in the direction of sustainability, and that again could change daily life in a way that provides even stronger support for sustainability.

To understand what sustainable rhythms are, it is necessary to understand their opposite, namely unsustainable rhythms. It is necessary to describe the restricted and unsustainable rhythms of daily life.

Restricted Rhythms of Daily Life

Most of us are living quite unsustainable daily lives. Let us take an example that most people in the highly developed part of the world know more than less. A type of family where the quality of life is strongly related to the quantity of consumption. They work to get the maximum amount of money which they use to maximise consumption. Because of that they use a lot of time driving in their car to different discount markets with good parking opportunities, and they always go for the cheapest offers. The diversity they meet in the shopping mall is strictly planned and controlled by multinational retail companies.

This restricted behaviour in the consumption system (the family) creates a pressure on the retail sector – they have to rationalise their own services, and they have to lay pressure on the suppliers to make the products cheaper. One of the supplier groups, the farmers, reacts to that pressure by intensifying the use of their land. Almost all land is gradually occupied by rational farming, supported by pesticides and fertiliser. That kind of farming reduces biodiversity dramatically, it pollutes the streams and rivers, and it affects the groundwater.

All the problems related to sustainability come from the social orientation of the consumption unit (the family). Or do they? Are not the family's consumption habits created by a controlled and restricted working life, where performance expressed in money is the only thing of value? And are not the consumption habits of the family created in a local community, strictly planned and controlled in every detail, where the work in the household is standardised, the local community is nothing more than the shopping mall, including commercialised entertainment. This picture is the rationality of daily life described by Ritzer in his famous book about McDonaldization (1996).

It seems as if daily life is locked into a rationality in which companies are under pressure from consumers which results in a rationalised and performance-orientated working life, and planned and controlled local communities that in turn results in a quantitative orientation in consumption which creates a further pressure on the companies ... The system is not rational but functionally coherent, and the system cannot by itself create a new functional rationality – for instance, related to sustainability. The system is locked. The only hope seems to be a strong (wo)man coming from somewhere outside who takes the necessary power to put forward an ecological reason and pacify all who oppose such a reason. Our only hope seems to be what Gorz called 'eco-fascism' (Gorz 1980).

Or is it really so bad? We could ask: If it really is so bad, why do living social and natural systems still exist? There must be something in daily life that allows people to arrange a street party, even though the planner of the suburb did not plan it. There must be something in daily life that makes it attractive, at least for some people, to buy organic food. There must be something in working life that sometimes gets workers to improve the quality of their products and services even though they are not asked to do that. In the rhythms of daily life there must be some openings for creating new connections, new insights and new aspirations. I wish to identify these rhythms.

Living Rhythms: An Opening

Sustainability is about regeneration and the development of social, natural and economic resources. Sustainability is about the creation of equilibrium and balance that makes regeneration and development possible. However, equilibrium and balance are not stationary, but rather rhythms that work together. The tightrope dancer can only keep her balance as long as she is in an alert rhythmical movement.

Balance in life is achieved through a rhythmic interaction between performance and enjoyment, and between exertion and rest. Balance in nature is achieved through a rhythmic interaction between the various rhythms of the species. The problems of sustainability are most likely rooted in the fact that the rhythmical interplay is not working, that rhythms have been disconnected from each other and have therefore begun to work against each other.

All living processes are rhythmic in nature. The seed sprouts in the ground in spring, sets the first leaves, flowers, sets seed and withers, only to repeat the same rhythmic process the following year. Human lives are also rhythmic, although the element of repetition from generation to generation is less than for the flower. The rhythm of life runs from childhood to youth, through the adult years to old age. The year has its rhythmic cycle of seasons, festivals, holidays and working periods. The week has a rhythm of working days and days off. Each weekday has a rhythm of sleep, waking up, travel to work, work, lunch, work, travel, household chores, leisure and sleep. Weekends have a slightly different and more individual rhythm. Being in a rhythm is to live. Being outside of rhythm can threaten one's identity and life.

Many individual rhythms are temporally and functionally connected and coordinated. This coordination used to be more directly linked to the movements of the sun. The rhythms of nature and humanity were linked together, and both social and natural rhythms were controlled by the sun's rhythmic movements (or more accurately, the movements of the Earth in relation to the sun). The sun, with its daily and annual rhythms, was the natural conductor of the rhythms of everyday life. This was very clear in the 'primary industries' (agriculture and fishing), to which the large majority of the population used to be linked. The fisherman had to follow the movements of the weather and the tides. He had to adjust his catch according to the various annual rhythms of the fish species. For the farmer, life was largely linked to the seasons, weather and the rhythms of the animals.

However, one of the great conquests of modern technology has been that it has made us – partially – independent of nature's rhythms. Employment in 'primary industries' has become vanishingly small due to huge rationalisations, and these industries have sought to break loose from the cycles of nature: heated and lit greenhouses and livestock buildings have been built so that vegetables can be grown and chickens hatched all year round. Annual fish migrations have been halted by constructing ocean farms.

For the vast majority of the population, who do not have direct contact with the 'primary industries', daily life is almost completely removed from the rhythms of nature. We have evaded the weather's oscillations by creating our own indoor climate in temperature controlled houses and vehicles. Electric light makes us no longer dependent on sunlight. You can buy strawberries in the supermarket all year round. Chickens are not born at Easter time, but all year round. We like to be able to wear summer clothes all year round, eat the same food all year round. Travel with the same ease all year round, shop any time of day, any day of the year. We can work at all times of day and, in some sectors, in any location. The rhythms of work have been split up and compartmentalised through work and time studies. This decoupling of rhythms has made us (partially) independent of time and place.

Rhythms have been detached from each other with the aim of optimising and rationalising each function separately. Rhythms have been reduced to controlled repetition. Modern technology has been the most significant means of achieving this fragmentation, but modern scientific forms of organisation have also contributed.

Throughout this modernisation process, there has been a loss of rhythmic coordination and cohesion. This loss represents a serious ecological and social impact. 'Put briefly and simply, the social time regulators subject the biological regulations to ever more serious tests' (Eriksen 2004: 244). Social and environmental sustainability is basically about re-establishing greater rhythmical cohesion. Balance, both ecological and social, is coordinated and cohesive rhythms.

It is not be possible to return to the former nature-dependent society. Eco-romantic ideals of returning to local communities linked to nature cannot create a sustainable foundation for humanity today. We are too numerous for local communities to be able to provide the necessary foundation for life, and we have a social orientation that goes beyond the local community.

Instead we have to develop new sustainable rhythms in the modern industrialised and globalised society. Rhythms that unite the activities of everyday life with the rhythm of nature; rhythms that allow human and social regeneration and development; rhythms that bring work, production and consumption in rhythmic balance with nature; rhythms that allow individual regeneration and development, and regeneration of social institution such as families, workplaces, local communities, cultural institutions and traditions; rhythms that create a structure of everyday life, but also rhythms which each and every one who participates in the rhythm influences. Rhythms that are interconnected, where the one rhythm supports the other.

To develop this idea I will, in the following, draw on three different theories: (1) Gunderson and Holling (2002) who develop a theory about adaptive cycles and what could be called natural rhythms in the relation between nature and man. (2) Giddens's theory of structuration (Giddens 1984), which conceptualises a social system in which social institutions are created by repeating activities which are constantly adjusted to each other. Both Gunderson and Holling and Giddens provide inspiration to a rhythm analyses related to sustainability. They create an approach to analysing rhythms, and they are used as an approach for interpreting 'rhythm analyses' developed by (3) Henry Lefebvre (2004). For Lefebvre 'rhythm analyses' is an approach to analysing everyday life.

Rhythms in a Chaotic Relation Between Man and Nature

A new systems approach is developing. It understands systems as incalculable complex entities in which change is discontinuous, in which chaos and order exist at the same time. Systems are organising themselves in non-linear behaviour. Sustainable systems are adaptive and have the capacity to change themselves in a world dominated by uncertainty. I will present that new approach by referring to Gunderson and Holling's book *Panarchy: Understanding Transformations in Human and Natural Systems* (Gunderson and Holling 2002). Gunderson and Holling's, and the network they have created, main interest is in the management of regional ecosystems in which natural systems, social systems and economic systems interact. Regional ecosystems are, however, only their reference for creating a general theory for understanding natural systems, social systems and economic systems.

Both natural and social systems are under tremendous pressure because of one-sided exploitation. Gunderson and Holling argue that even planning and management for sustainability quite often create an unexpected pressure on natural and social systems. Many water management programmes, for instance, where water has been regulated in large controlled systems, have caused catastrophic flooding when man and nature have acted unforeseeably. Another example is pest control in the natural environment which has unbalanced the ecological balance. Against this background Gunderson and Holling ask a very fundamental question: How is it that many natural systems, social systems and economic systems have not collapsed in spite of one-sided exploitation of nature, one-sided exploitation of human resources, and economies being under constant pressure? The systems must have a very strong resilience to survive. This leads to the main argument: planning for sustainability should understand and recognise the resilience of the systems, and it should not aim to establish detailed control systems, but rather support the resilience of the systems.

Resilience, as Gunderson and Holling use the concept, is based on what I would call sustainable rhythms. Gunderson and Holling define resilience not as a fixed condition. Resilience is created in a moving system where a multiplicity of different rhythmic cycles interplay and adapt to each other.

Resilience is low in a fixed system. It has no ability to adjust itself to changing conditions. Also, a heavily controlled system has a low resilience because it is not able to act rationally in a complex situation, where reality looks different from different positions and where the context is changing constantly. A water system constructed by well-dimensioned drainage works can create disasters when the weather or the humans do not act as expected. 'Water systems' consisting of different biotopes, wetlands and constructions that are prepared for flooding are much better prepared for a situation where the unforeseen occurs.

A system with low internal variety or diversity has a low degree of resilience. A system with a limited variety in external connectivity also has a low degree of resilience. If a system has limited possibilities to draw on resources from other systems, it will have a limited resilience. An ecosystem with a low diversity of species is more vulnerable than an ecosystem with a high diversity. A local community dependent on only one industry is much more vulnerable than a community with a high degree of industrial diversity.

Most planning and management has striven for controlled repetition in what ideally has been seen as the most functional systems. That kind of planning has its strength in making every single function more effective. However, such planning reduces the resilience and, because of that, it actually creates a lack of sustainability, even when the purpose of the planning is to favour sustainability.

Gunderson and Holling suggest another principle for planning and management. In their view management should focus on persistence, adaptation, variability and be very much aware of unpredictability – anticipated uncertainty.

Gunderson and Holling find their main argument against traditional planning and management in what they call the adaptive life cycle. All living systems, human and natural, have a life cycle which cannot be predicted but can only be reconstructed afterwards. This life cycle has four phases: a beginning, where a newborn system starts to exploit some given opportunities and gradually creates itself, this is the flourishing phase. Then the system will come to the (second) phase of conservation. Here resources are stored and structures refined. But at a point the third phase will be reached: the system will face a serious crisis. Structures will fall apart and energy will be released. However, if the system has the capacity, it can turn out to be a creative destruction that leads to the fourth phase: that of reorganisation. The system reorganises itself and finds new opportunities to exploit, and then we are back to phase one.

According to Gunderson and Holling, an adaptive cycle that aggregates resources and periodically restructures to create opportunities for innovation is a fundamental unit for understanding complex systems from cells to ecosystems to societies to cultures. Such universal theories are, of course, always questionable. In this case, Gunderson and Holling have created a hypothesis, which is not testable. For instance, it will, in practice, be impossible to distinguish between incremental adaptation and more radical reorganisation. However, if we understand Gunderson and Holling's life cycle as a metaphor, and not as a hypothesis, it is more usable. It should help us to see the world not in mechanic and linear terms, but in organic and non-linear terms.

When we keep the adaptive life cycle in mind, we can see that managers and planners make a big mistake when they try to fix a system or, even worse, to turn it back. Instead they have to go with the system and contribute to the resilience of the system. Another important basic insight provided by the adaptive life cycle is that the conditions for creating resources, variety and learning are very different in the different phases of the cycle. What is right in one phase could very well be wrong in another.

Functional diversity maintains ecosystem resilience. The self-organisation of ecological systems establishes the arena for evolutionary change; the self-organisation of human institutional patterns establishes the arena for future sustainable opportunity.

Sustainability requires both change and persistence. It is maintained by relationships among a nested set of adaptive cycles arranged as a dynamic hierarchy in space and time. This is Gunderson and Holling's notion of 'panarchy': an ordered anarchy, containing millions of interrelated rhythmic cycles.

Social Systems Between Repetition and Change

There is a parallel between Gunderson and Holling's theory about order and anarchy in natural systems and in systems connecting man and nature, and Giddens's theory of the structuration of social systems or institutions (Giddens 1984). Giddens emphasises that institutions most of all consist of routine activities – I will call them rhythms. Usually, we use our practical consciousness when we are performing our daily rhythms. Most of what we are doing is carried out unreflectingly though this does not mean that we are functioning as mechanical instruments. We are continually using a kind of what Giddens calls monitoring reflexivity. That is, we are constantly aware of whether the rhythm is going as it is supposed to, and we are always ready to make slight adjustments to the rhythm to fit changes in external conditions. We are performing our rhythms with a sort of practical and adaptive consciousness. Sometimes we are also using our discursive and intellectual reflexivity, asking ourselves and those we are in rhythmic relation to: Could this be done in another way? This practical adjustment of the daily rhythms gradually changes the institutions and the systems, sometimes intentionally, but most often unintentionally. At one point, the many small incremental changes have changed the institutions so radically that it either collapses or reorganises itself.

There are clear similarities between Giddens's theory of structuration related to the social world and Gunderson and Holling's theory of panarchy which is mostly related to nature (though to some extent to social systems too). However, in the concept of sustainability, nature and society are interconnected. Rhythms of everyday life are performed in an interconnection between the rhythms of nature, the rhythms of man-made technologies, and the rhythms of social systems and institutions – rhythms with a more or less harmonic interplay in everyday life. In rhythm analyses, it is possible to analyse social conditions, technology, artefacts and nature together. It is possible to cross boundaries of disciplines and sectors. Rhythm analyses emphasises everyday life in the analyses of changes in the direction of sustainability. The idea is neither to construct a new system from outside, nor just to understand the system as it is. The idea is to find principles for the rhythm of human life that allow living subjects to gradually create a more sustainable world. The philosophy of Henry Lefebvre can help us to define such principles. There are three key concepts in Lefebvre's philosophy, which I will make use of to form principles for sustainable rhythms. The concepts are: repetition, discontinuity, and Lefebvre's concept of rhythm, which is constructed out of the two former concepts.

An Analytical Concept of Rhythms: Repetition, Discontinuity and Rhythm

Lefebvre was a philosopher who had an interest in everyday life. For him rhythm and rhythm analysis (Lefebvre 2004/1992) are useful keys with which to produce a critical understanding of everyday life. Rhythms structure everyday life. Rhythms connect people to places, connect and disconnect people in time and space. And, I will add, everyday life meets the nature in a rhythmic relation, and our relations to nature are maintained and recreated through rhythms.

Rhythms are, according to Lefebvre, created in a dialectic relation between opposite phenomena: rhythms are created by repetition AND discontinuity. Rhythms are continuity AND discontinuity. If we live in a world of discontinuity, all social and natural systems would collapse. If we, on the other hand, live in a world of continual repetition, the spirit of life would waste away.

Rhythms are re-creations, in time and space, created in an interplay and interconnected with many different autonomous rhythms. The day and the year have their rhythms which are in part created by the repeated movements of the sun and the earth. The rhythms of the day are, however, very different from the rhythms of the light coming from the street lamps, such light is mechanical and repetitive with no differentiation or recreation. The rise of the sun is also a result of mechanical forces, but the rhythm of the day is not only created by the sun. The rhythm of the day is made in interaction between the sun, the atmosphere, the sea, birds and animals. Because of the steady repetition and differentiation, every dawn has its unique charm.

Lefebvre's point is that the element of repetition is too dominant in modern life. The spread of capitalism, the spread of modern technology (industrialisation) and the spread of bureaucratic control have favoured repetition in daily life, and have reduced our relation to nature to repetitive exploitation.

Repetition is doing or almost doing exactly the same again and again, and consequently there is by definition no creativity. Repetition is brutal, tiring, exhausting and tedious (Lefebvre 2004: 73). However, repetition is a type of activity that can be very effective in achieving a certain goal. The repetitive movements of the slaves in the Roman navy created effective warships. The repetitive and controlled activities of the workers in the mass producing industry created a very high degree of work efficiency. However, they also created alienation, the destruction of human capacities, and the destruction of health and a lack of social sustainability. The planning and structuring of the local communities created a high degree of predictability and a high degree of efficiency of consumption in family life. However, it also created futile suburbs, social degradation, crime, loneliness and a lack of social sustainability. The repetitive growing of the same crops on the same land, undisturbed by other species by the regular use of pesticides and fertiliser, creates a high yield. However, it also creates a less fertile earth. Pesticides are spread throughout nature, pollute the groundwater and the streams, they produce foodstuffs with less nutrition, and lack both environmental and social sustainability. The worshipping of repetition has made each single function very effective. It has, however, also created a lack of sustainability and consequently restricted productivity because of the misuse of social and natural resources.

The worshipping of the one aspect of the living rhythms – repetition – is, according to Lefebvre, a crucial critical aspect of modern society. However, over the last decade it has also been quite clear that the one-sided worshipping of the other aspect of living rhythms, differentiation and discontinuity, can be just as fatal. As Sennett (1999) and others have pointed out, the widespread flexibility found in modern life, with little stability and repetition, is a threat to personal integrity, to binding social relations and to value systems. If flexibility and discontinuity dominate, it will be impossible to find the social resources to create sustainability.

Repetition in Daily Life

The expansion of controlled repetition has characterised the twentieth century, and the trend has continued into the twenty-first century. Work systems have been marked by 'scientific management', work is studied scientifically, new standardised and controllable work tasks have been developed, and every work movement is given a standard time. Small producers have been replaced by big units of mass production, controlling production and consumption. The exploitation of the land has been one-sided, repetitive and controlled by pesticides and fertiliser. Daily life is increasingly regulated by rules and decreasingly by direct social relations. In the shopping malls, entertainment centres and traffic systems we are shepherded like sheep.

This society has abolished unlimited poverty (in the rich part of the world) but it has, at the same time, developed in an unsustainable direction. Human and natural resources are destroyed on a massive basis. Repetitive and controlled work destroys the workers' ability to learn and to develop. It is well documented that repetitive and controlled work is harmful for both physical and mental health (Karasek and Theorell 1989). Local human resources are undermined by bureaucratic management and centralised, planned systems. The fertility of the land and biodiversity are threatened by the one-sided, repetitive and controlled use of the land. Mass production is linked to mass consumption with an increasing use of non-renewable natural resources, one could continue.

Social scientists working in the 1980s and 1990s termed this repetitive and controlled form of society as 'Fordist' (Boyer and Durad 1997) which they were convinced would gradually be replaced by a post-Fordist society. New market conditions, new technologies and the undermining of national economies by globalisation would gradually replace the notion of repetition with the notion of flexibility. However, the whole idea of Fordism and post-Fordism is now almost forgotten. The reason probably is that repetition and control is still very strong in the twenty-first century. 'Scientific management' has been refined as is reflected in the concept of 'lean production'. Quality control and environmental management have developed as new, refined control systems. IT systems have invaded daily life with their sophisticated control. However, the new control systems seem to be more internalised than those of the 'classic' Fordistic society. Companies employing the concept of lean production invite the employees to participate in the development of an even more repetitive and controlled production system. When environmental management is implemented, the employees are invited to participate. The idea is, when the system has the opportunity to exploit the practical knowledge of the workers, it can be even more repetitive and controlled. When the system is running, the employees are supposed to report environmental data to the system according to formalised procedures. We see the same trend in the development and implementation of environmental systems outside production, for instance environmental systems managed by public authorities.

There is a possible objection to this line of argument, namely that continuous improvement has, for some years now, been a guideline for quality systems, including environmental management systems. However, it is a way in which to gradually refine repetitive control and it seldom releases human resources. The internalisation of the control systems makes it even more difficult to escape this control, to cheat the system as it were. Repetitive control becomes more advanced.

However, no system can survive without people performing some kind of monitoring reflection as Giddens made clear, we can ask: Are things going on as they are supposed to do? Is the noise, the smell, the feeling right? Are we running into something unexpected? Without people undertaking this monitoring reflection, systems will very quickly collapse. If we all work according to the rules, everything will stop. Sometimes this monitoring reflection turns into a discursive reflection: When the producer of service products meets her customers, they will discuss quality, they will find ways to cheat the system, thus creating small innovations. Neighbours will talk and find ways to help each other outside the formal economic system, etc. Despite the dominance of the repetitive element in modern society, living rhythms are re-created all the time, which gives even a restrictive system a certain degree of resilience, and creates opportunities for sustainable development.

Discontinuity in Daily Life

Our time is marked by imposed, controlled repetition. The opposite is, however, also a reality – differentiation and discontinuity constitute a strong tendency in modern life.

New constantly turbulent global market conditions and new technologies that to some extent release us from bindings to time and space erode continuity and in doing so destroy vibrant rhythms.

For a considerable part of the working population, work has become borderless (Lund and Hvid 2007). We can work at any time, and spare time is residual time. We do not have colleagues, but only temporary co-workers. In so-called spare time there is no room for binding social relations because one never knows when one is available. These are conditions which put a serious strain on established kinships, make engagement in activities in the local communities impossible and affect family life negatively.

Furthermore, the organisation of everyday life outside the workplace is moving towards discontinuity. The time structures of cities are vanishing. All modern cities are moving in the direction of New York, the city that 'never sleeps' as Frank Sinatra sang. Fast food is a threat to the home-made meal, and thus a threat to the time structure of the family. Cheap and effective transport systems make it possible for many to live at different places at the same time.

A high degree of discontinuity is connected to a high degree of mobility. Mobility is a serious environmental problem because mobility creates transport and transport creates noise, destroys local areas and creates serious emission of greenhouse gasses.

However, it seems as if some structures sometimes create themselves out of this discontinuous chaos. People are creating networks among common interests. Steady kinships are still created. Family life still exists. The number of 'one issue movements', related to environment, health, local issues, culture, etc. is growing. Also, sustainable rhythms can develop in a social setting of discontinuity.

Sustainable Rhythms

Rhythms are shaped by two different components, repetition on the one hand and discontinuity and differentiation on the other. In modern society both components are strengthened, as it was argued above, but most often without creating rhythms. Modern everyday life contains different and separated spheres. Some can be characterised by brutal repetition, and others can be characterised by discontinuity. The big challenge of sustainability is to unite these different spheres into a coherent everyday life having sustainable rhythms, uniting repetition and discontinuity. A simple and constructed example can illustrate the point. McDonald's, as depicted in Ritzer's books about McDonaldization (1996, 2002), has been a metaphor for controlled repetition in modern society.

First Version of the Story

Peter from Copenhagen goes for a trip around the world, like many other young people of his age. His life has been heavily controlled by the school system for many years and now he wants to release himself from that pressure. He wants total discontinuity. He only wants to respect the time schedules of the air companies that will take him round the world. However, Peter does not have much money but for some years he has been working at McDonald's in his spare time and now he can get a job in most cities in a McDonald's restaurant. So, in every big city the plane takes him, he starts by visiting all McDonald's restaurants until he has a job. If there are no jobs, he goes on to the next city. When he has been working for two weeks, he takes a one-week holiday. After that, he goes on to the next city. When Peter comes back to Copenhagen after one year he has learned one thing: to combine discontinuity and repetition without creating rhythms. Peter has learned to live under discontinuous conditions in different places and with new people and with no binding relations to anyone. And he has learned to use the globalised repetitive structures, in this case McDonald's, to maintain the discontinuous and disconnected living. Peter has not travelled in the direction of sustainability. This was the realistic version of the story.

Second Version of the Story

Peter does as he planned to do. In Kuala Lumpur he works with Aisha. They are both working backstage. She is new. He is trained. He teaches her how to do things. She learns quickly. One day they decide to go out to have a cup of tea in another restaurant. Aisha is studying sociology. She has read *The McDonaldization of Society* by Ritzer and she explained the idea to him, and they discussed how it fits with their own experiences. After some days, Peter continues his travels but they keep in contact via e-mail, and they continue their discussions. One year later Aisha comes to Copenhagen. Now they are ready to act and a global movement against McDonaldization is born. (What else happened between Peter and Aisha is not relevant here. This is not a Hollywood love story).

This was the naïve version of the story, but it is in line with current optimistic views about globalisation. One exponent for that point of view is Giddens who has a positive view about reflexivity in late modernity (Giddens 1990). Peter and Aisha are both 'dis-embedded', in Giddens's terms, from any binding social relations. They are free, including the freedom to make their own reflections, inspired by books, movies and the mass media. They are free to create a new version of David against Goliath: Peter and Aisha against McDonald's. However, their story is competing with thousands of other stories, for instance those about how to be rich, how to be slim, how to achieve a happy marriage. And it is quite difficult for Peter and Aisha's disconnected story to compete with these other stories related to individual lives.

Third Version of the Story

Peter takes the plane to the first city but there is no work at McDonald's. He takes the plane to the next. No work, and so it continues. McDonald's is losing market shares. It has to close down restaurants. So, after two weeks he must go back to Copenhagen, now without any money. The day after arrival he has to go to McDonald's. Here they are very glad to see Peter. They need experienced employees who know the local communities and the local customers. McDonald's has changed its business concept. To gain market shares all restaurants must adjust themselves to the local market. McDonald's still offers a restaurant with all that belongs to it. McDonald's offers financing, accounting, etc. But each restaurant has to find its own identity. In Peter's restaurant they have a meeting among the employees asking themselves: What can McDonald's do for the local community? Afterwards they have meetings with the local environmental groups, the schools, local workplaces and others, and gradually McDonald's turns out to be a living part of the local community. Peter has not jetted around the world, but he has participated in the creation of sustainable rhythms in his own community, created in a multinational setting. Rhythms created in the combination of the repetitive structure of McDonald's and the creation of local specificity. It has changed his life, it has changed McDonald's and it has changed the local community.

This was the utopian version of the story, but is it perhaps also another version of the realistic story? Actually, it has to be realistic because here we are talking about sustainability. We cannot survive without it!

At least there are many realistic elements in the third story. McDonald's is losing its market share and it is looking for new business concepts. Product diversity, mobilising the competences of the employees, involving the stakeholders are not new ideas just created in this short story. They are well-established buzz words in the business world. Local authorities involve themselves in the development of the neighbourhood. Union representatives would love to involve themselves in projects like this. Local authorities' collaboration with local business is not unrealistic either. Environmental organisations working together with business are now relatively common. None of the above-mentioned organisations, institutions and authorities can create sustainability. No single institution or organisation can. Sustainability must be created in collaboration!

Concluding Remarks

In this chapter 'rhythms' is used as a conceptualisation of daily practices in everyday life. Rhythms are seen as repeated activities structured in time and space. Rhythms are routinised practices in our material and social lives – in households, in local communities, in the transport system, in work places and in our practical relations to nature. These rhythms are related to each other in a balanced or in a conflicting manner.

The main point of the chapter is that sustainable rhythms in everyday life, as they are characterised in the beginning of this chapter, are a decisive factor in the creation of a sustainable society. At the same time, the concept of rhythm creates an opportunity to develop a common frame for research and practices to improve the sustainability of everyday life. If we look for the rhythmic aspects in different spheres of daily life, we will find opportunities to improve the sustainability of daily life and to connect different spheres of life. Here I will give some examples:

Research and practice related to the development of regional ecosystems is one source for detecting sustainable rhythms. Here I have already referred to Gunderson and Holling (2002). I could also refer to my colleagues Birger Steen Nielsen and Kurt Aagaard Nielsen (2006) who analysed popular involvement in the local development of ecosystems in an action research perspective, and emphasise the importance of the sensuous, the aesthetic and the patterns of day-to-day relations between man and nature.

In the literature and practice related to learning we also find support and inspiration for sustainable rhythms. Here I will especially mention the work of

Wenger and those who are influenced by him (Wenger 1998, Wenger, McDermott and Snyder 2002). They argue that innovation in an organisation and learning in a society is strongly related to the learning capacity in existing communities of practice. A high capacity for learning needs rhythms in the community which makes formal and informal communication and interaction in the community possible, and which makes formal and informal communication and interaction with related communities of practice possible.

I also find support and inspiration for the concept of sustainable rhythms in the Scandinavian literature about the development of work systems (Thorsrud and Emery 1969, Docherty, Forslin and Shani 2002, Hvid 2006) and the work reform movement to which this literature is related. For decades this movement has fought the brutal repetition of Taylorism and bureaucracy. The alternative has, however, not been the individualised and discontinued work systems. The alternative has been autonomic collectives of producers installing rhythms in daily work life based on competences and preferences and, of course, the rhythms of production, suppliers and users.

References

Boyer, R. and Durad, J.-P. 1997. *After Fordism*. Basingstoke: Macmillan.

Docherty, P., Forslin, J. and (Rami) Shani, A.B. 2002. *Creating Sustainable Work Systems: Emerging Perspectives and Practice*. London: Routledge.

Eriksen, T.B. 2004. *Tidens Historie*. København: Tiderne Skifter.

Giddens, A. 1984. *The Constitution of Society*. Cambridge: Polity Press.

Giddens, A. 1990. *The Consequences of Modernity*. Cambridge: Polity Press.

Gorz, A. 1980. *Ecology as Politics*. Cambridge, MA: South End Press.

Gunderson, L.H. and Holling, C.S. 2002. *Panarchy: Understanding Transformations in Human and Natural Systems*. Washington, DC: Island Press.

Hvid, H. 2006. *Arbejde og Bæredygtighed*. Copenhagen: Frydenlund.

Karasek, R. and Theorell, T. 1989. *Healthy Work: Stress, Productivity, and the Reconstruction of Working Life*. New York: Basic Books.

Lefebvre, H. 2004. *Rhythmanalyses: Space, Time and Everyday Life*. London: Continuum.

Lund, H. and Hvid, H. 2007. *Øje på det Psykiske Arbejdsmiljø i Grænseløst Arbejde*. Copenhagen: Landsorganisationen Danmark.

Nielsen, B.S. and Nielsen, K.A. 2006. *En Menneskelig Natur. Aktionsforskning for Bæredygtighed og Politisk Kultur*. Copenhagen: Frydenlund.

Ritzer, G. 1996. *The McDonaldization of Society: An Investigation into the Changing Character of Contemporary Social Life*. London: Sage Publications.

Ritzer, G. 2002. *McDonaldization*. London: Sage Publications.

Sennett, R. 1998. *The Corrosion of Character: The Personal Consequences of Work in the New Capitalism*. New York: W.W. Norton and Co.

Thorsrud, E. and Emery, F. 1976. *Democracy at Work: The Report of the Norwegian Industrial Democracy Program*. Leiden: Martinus Nijhoff Social Sciences Division.

Wenger, E. 1998. *Communities of Practice*. Cambridge: Cambridge University Press.

Wenger, E., McDermott, E. and Snyder, W.M. 2002. *Cultivating Communities of Practice*: *A Guide to Managing Knowledge*. Boston: Harvard Business School Press.

Chapter 2
A Record on Modernity, Rationality and Sustainability

Bo Elling

Modernity and Rationality

Modernisation

In my view, the point of departure for the discussion on sustainability must be taken in a sociological concept of modernisation. A comprehensive outline of this concept is not attempted here; I limit my exposition to the main points.[1] Modernisation is the process that transforms society from being reproduced by cultural traditions, norms and habits to a society characterised by reflexivity and a societal divide into autonomous parts, fractions or sectors. Thus, the concept involves clarifying how modernity differs from traditional societies and, moreover, how the freedom to act is achieved within modernity. Habermas's basic metaphors for such a societal divide are 'systems' and the 'lifeworld', in which different action orientations and their related rationalities exist. His term for this process is out-differentiation, which results in the difference between actions oriented towards success and actions oriented towards reaching mutual understanding.

Action Orientations and Different Types of Rationality

The outcome of modernisation is the realisation of a society without a centre for a common will and reason to control development, a centre which used to constitute the core and symbolic unit of society. Modernity also unleashes a series of fundamental *contradictions* between the various autonomous parts of society (systems) now functioning on their own respective terms and pursuing their respective, exclusive, objectives. Such goal orientation is also what creates the continued historical dynamics of society, and lays the groundwork for basic conflicts and dysfunctions in society.

Habermas speaks about instrumental action which is related to the orientation of action described above. Instrumental action is conducted in systems and relates to systemic rationality which Habermas terms *cognitive-instrumental rationality*.

---

1   A comprehensive account for a concept of modernisation is given in Elling (2008: 69–169).

Such actions focus on systemic goals and requirements without considering the terms and needs of society as a whole. The criterion for rationality in instrumental actions is *efficiency*, i.e. achieving systemic objectives as far as possible with appropriate means.

In addition to cognitive-instrumental rationality, ethical and aesthetic rationalities are out-differentiated in relation to systemic action within the *expert systems* of morality and art. Norm-regulated action within the moral expert system is validated in relation to *moral-practical rationality* and measured by *rightness* (in following the norm). Dramaturgical action within the expert system of art is validated in relation to *aesthetic-expressive rationality* and measured by truthfulness (in saying or expressing what is really meant or intended).

As *integrated* criteria for rationality (i.e. opposed to out*differentiated*) cognitive-instrumental, moral-practical and aesthetic-expressive rationalities exist solely in the lifeworld, termed *communicative rationality*. As soon as moral-practical and aesthetic-expressive rationalities are laid down as criteria for action within a systemic context, they are applied with a certain pre-given purpose within the system and thus instrumentalised for the fulfilment of that purpose. Consequently, they can only be reintroduced into systemic actions as a claim from the lifeworld.

Substance and Value

The divide between a useful substance and an economic value is another scientific metaphor for the division of society in the modern period. This division is associated with commodities and resources in the capitalist economic system and characterises the duality of something material (utility value) and a social structured economic value (exchange value). These two sides or aspects cannot be separated. They are, however, often separated in theory, when, for example, industrialism and capitalism are viewed to be two independent parts of modern society.[2] Another example is when sustainability is understood as something purely material or technical without taking into account economic values and their social structuration. In making such separations, everything is possible in theory but not in reality. Making such separations can be compared to ignoring market realities in a capitalist society when claiming that technical matters alone can define the options for action.

Utopia or Illusion

In short, it is the utopia expressed in the concept of sustainability that will reunify all three criteria for rationality – also in systemic actions – first of all within the economy and within policy or one should in fact say 'by means of policy'. Another metaphor could be that the concept of sustainability will let utility values control

---

2   This is, for example, what Giddens does in his description of the four dimensions of modernity; see Giddens (1990: 55–6).

our actions instead of the market – say, economic values. The core question is whether this will be possible in reality – locally, regionally and (not at least) globally. Mainly because as it will be made evident below, the utopia of the concept of sustainability described here should rather be seen as an illusion.

The Concept of Sustainability and its Contradictions

Utilitarianism and Kantian Ethics

The concept of sustainability lays a strong emphasis on ethics. According to the main prescription of the concept we must act for the sake of future generations and we must act in relation to something more superior than ourselves – nature. This is a moral claim that relates to the modern divide between Kantianism (duty-ethics) and utilitarianism (utility-ethics). Sometimes Kantianism is also called 'temper-ethics', i.e. in which temper an act is undertaken. Unlike consequentialism that maintains that the consequences of an act are the basis for deciding whether an act must be considered good or bad.

Kantianism is *not* to be confused with an act that must be performed because of societal norms. On the contrary, duty comes into the picture in what oneself considers being a good moral action. This is the famous categorical imperative in Kant's work. According to Kant we must deny actions directed to us from authorities if we ourselves consider such actions to be unmoral – whatever consequences such a denial will result in for ourselves. This is equivalent to the modern *emancipation of the individual*. It is the quintessence of the process of modernisation. The modern *subject* is born in that process and given autonomy in relation to such authorities that in pre-modern societies ruled individual action. We ourselves as individuals, and *not* formal authorities or authorities as they appear in habits, traditions and culture, can define the norm. Kant termed these circumstances the *autonomy of the subject* and for him it was the mode of modern life.

The moral claim that we should act to achieve sustainability in all we do is obviously based on duty. However, it is based on utility as well. Its utility goal orientation towards economic, social, environmental and generative equity contrasts with its duty orientation towards protecting nature with all our actions no matter if such actions are of benefit to ourselves or not. Put another way, an action can be for the benefit of one or more individuals or even mankind, but if it is not for the benefit of nature it cannot fulfil the duty side of the moral claim. Saying that the protection of nature is in the end for the benefit of mankind does not solve the contradiction between the two ethical claims since this will make the protection of nature into something goal-oriented for the benefit of mankind and not for nature in itself. In fact, all anthropocentric moral statements will contain such inherent contradictions. Making contrasting statements of an ecocentric moral character will not solve the problem since it will just neglect the utilitarian claims about sustainability.

Having said this, it should be obvious that the two ethical aspects of the moral claim for sustainability cannot be fulfilled at the same time. So, it should be asked if we can develop the concept so that contradictions of this kind can be avoided.

Autonomy, Intersubjectivity and Social Recognition

However, the concept of sustainability *can* be said to go beyond a Kantian focus on the autonomy of the individual. It will commit the individual as well as the whole society to objectives that go beyond or are guarded by something more meaningful than the individual or society. This cannot *simply* be *nature* or *mankind,* as already demonstrated above;[3] it must be something that *also* goes beyond the existing *societal* reality. Though Kant can help us to identify the contradictions in the moral claims of sustainability, his categorical imperative is not adequate to establish modern communities where common norms and values are shared, as Habermas has demonstrated.[4] Habermas's answer to this inadequacy is to replace the centrality of the autonomy of the subject with the notion of *intersubjectivity.* Only such norms and statements that we can establish in reaching mutual understanding can survive as community shared norms. Axel Honneth (1995) takes the argument one step further than Habermas in pointing out that it is not positively formulated principles concerning moral claims that establish a norm, but the violation of intuitively given understandings of justice. This means that *social recognition* is the normative requirement for all communicative action. Habermas finds it necessary to replace the autonomy of the subject and argues that we can only establish such norms that we can agree upon in mutual understanding (intersubjectively), and Honneth wants to highlight that we cannot survive as individuals unless we can experience social recognition. That is why we in moral matters must act consistent with our own beliefs as well as what is valued by other humans. This is not always an easy task.

Environmental Standards and Responsibility for our Actions

A common understanding of the concept of sustainability tells us to act in relation to nature's capacity. This basically contradicts the concept of modernity in which it is society's own concepts of rationality that must be the point of departure for appropriate actions.

In traditional pre-modern societies, on the contrary, *culture* frames actions. In this context framing should be understood as being ruled by norms, habits and cultural traditions that stem from all parts of society in common. Such framing also implies an 'adaptation' or 'balance' with nature's capacity. So culture is opposed to modernity in which systemic dynamics define goals and means related to systems' needs without taking the surrounding society into account, including human and natural capacity.

---

3   A longer account of this can be found in Elling (2008: 15–67, 69–169).

4   Habermas (1984: 273–337), Part III, Intermediate Reflections: Social Action, Purposive Activity, and Communication.

At the end of the day, in a modern society, it is a *social* reality that, for example, human capacity in industrial work is solely challenged or limited by the efficiency of its output. Likewise, an ecosystem's capacity will be laid down by its utility for economic employment. If more than one system uses the ecosystem, conflicts will appear over the use, not what the ecosystem can withstand.

The answer to this problem cannot be to re-establish nature as the limit. Instead modern society has to define its own limits by going beyond system limitations and strive for balancing needs and objectives in terms of society as a whole. This means balancing systemic rationalities with rationalities from the lifeworld.

In other words, if we let nature set the limits, we introduce a new authority. We believe that it is possible to act in a pre-modern fashion and ignore our responsibility. We leave it to nature to tell us what to do. On the contrary, we must create 'balance' in our social actions by introducing such criteria for rationality that we can control ourselves. We have to supplement systemic criteria for rationality with criteria from the lifeworld and not from nature.

The answer to the problem of how to guide social actions towards sustainability can neither be to introduce a criterion for sustainability concerning social economic equity, since the necessity to achieve economic profits, i.e. maximal economic efficiency concerning the outcome of an action, still exists for the specific action.

The Concept of Sustainability is Systemic

In what I have said above I have argued that the concept of sustainability may be seen as a utopian idea, and alternatively as an illusion. But for the most it must be considered to be systemic. For example both in the Brundtland Commission's Report and within the World Bank, economic, social, legal and environmental justice and equity, and equity between generations are core issues. The concept is systemic because it primarily aims to ensure the conditions for the continued existence of the modern economic system in which the negative impacts of the contradictions within this system – between different participants (socially and economic) and between future participants (future generations) – are abolished or compensated. The concept in this form is goal-oriented and furthermore does not include reflections on how these contradictions can be understood and abolished by employing another economic model.[5]

Contested Natures

In their book *Contested Natures*, Macnaghten and Urry (1999) argue for a rethinking of the relation between nature and society. We must stop considering

---

5   I give an account on basic differences between a goal-oriented process and a process oriented towards reaching understanding in Elling (2008: 171–218).

nature as something 'out there', as if there were a realm of nature separate and distinct from that of culture (Macnaghten and Urry 1999: 16).[6]

In other words, they reject the idea that the environment can be regarded as a 'real entity', which, in and of itself and substantially separate from social practices and human experience, has the power to produce unambiguous, observable and rectifiable outcomes (Ibid.: 1). Such a strategy they call 'environmental realism', thus they deny that nature should be an object for science in a strict sense. But they also reject what can be termed 'environmental idealism', in which nature is considered in terms of values; this approach claims that we can justify certain values for the social because these values are inherent in nature.

Finally, Macnaghten and Urry refer to so-called 'environmental instrumentalism' which explains appropriate human motivation to engage in environmentally sustainable practices and hence the resulting environmental goods and bads (Ibid.: 1). But such thinking must also be rejected since people's relationship to nature is ambivalent and their actions and beliefs depend more on their beliefs and relations to other people and the specific context than on public or authoritative demands or inducements. In contrast to all three -isms they ague for an understanding of nature as a diversity of contested natures, where each such nature is constituted through a variety of socio-cultural processes from which such natures cannot be plausible separated.

Seen in relation to the concept on modernity described above, this rethinking is problematic. Indeed, it can highlight that most of what is normally termed nature is in reality culture, i.e. cultivated or manufactured nature, in contrast to what can be termed original nature.

But Macnaghten's and Urry's concept of socially contested natures can easily draw a veil over the fact that such societal processes cannot take place without a natural basis. In addition, a natural basis that those societal processes are in conflict with since they are destructive and not just constructive. The natural basis should not purely be understood as 'a place to stay' or 'a place to take place', such as in classical economic theory. Furthermore, it is a substance or material to work on, a natural law for exploitation, for example in chemical or physical processes. Finally, one can think of emissions from societal processes of production, for instance CO_2 that will affect the natural basis, such as in the case of climate change.

That the climate is a social construct in the sense that modern society greatly affects the climate and its development is not the same as if society can 'control' the development of the climate. Not that we can ignore the climate as a natural basis for the modern economy, as it also was for traditional economies. We cannot decode those laws of nature that are part of climate change so that they will work in accordance with modern societal dynamics. In other words, that we can have an *impact on* something is not the same as we can *control* it.

---

6   Many others have pointed to this problematic. See, for example, the numerous and highly relevant articles in Lash, Szerszynski and Wynne (1996).

In stating that social processes form chance nature to culture Macnaghten and Urry ignore that, in particular, those social processes which, as they themselves point out are dynamic, are themselves in the modern epoch outdifferentiated from cultures and form their own life, i.e. are developing on their own premises and rationality and in doing so are in conflict with the natural basis. By ignoring this, they blur the contradiction between the social and nature, which in fact is not blurred by social construction, but on the contrary is caused by the ongoing instrumentalisation of the social.

Ecological Modernisation

Something similar can be said about the neo-modernistic trends within sociology, when they claim that sustainable development can be achieved by an ecological modernisation of society. Though such authors highlight the social construction of nature, they either stay within the framework of an independent economic analysis or they jump totally out of such a framework and claim the possibility of forming nature as we want it.

An example of the former is Arthur Mol's concept of economic and ecological rationalities that can be put on a par when undertaking commercial activities (Mol 1995). Thus, he stays within a systemic framework of understanding. It can also be said that he 'imports' the natural basis into the systemic economy and by doing so he removes the contradiction between those two. Consequently he has severe difficulties in identifying ecological rationality, which he cannot define in a wider sense than that the business companies internalise environmental costs to the extent that it can benefit their position on the market.

An example of the latter is Maarten Hajer who claims that we should not discuss how to protect or preserve a certain part of nature, but quite the opposite, we should pose the question of what nature and what society we want (Hajer 1996: 259). Thus he claims the possibility of forming both nature and society in a way that we find appropriate and in doing so ignores the basic contradictions between systemic productivity and a given natural basis.

In both cases sustainability becomes a postulate. In the first case it is claimed that the independent economy can go beyond the framework of independence and include external considerations. Thus it is claimed that the objectives of the economic system can be achieved on the basis of ecological premises that can be injected into the economic system from outside. The second case becomes a postulate because it claims that it is possible in general to ignore the objectives of the systemic economy and create a nature – and a society too – that is not marked by the objectives of such an economic system.

In both cases we are back to a situation in which it is claimed that if we ignore the capitalist formation of social processes we can imagine everything, both within the social-economic process itself and what concerns shaping a nature from ecological objectives as well. However, as demonstrated, these claims remain pure imagination.

In their article 'Ecological Modernisation and Institutional Transformations in the Danish Textile Industry' Søndergaard, Hansen and Holm (2004) provide examples of how the modern Danish textile industry, during the 1990s, was very capable of developing different ways of production in which ecological considerations were taken into account.

However, the authors also demonstrated that the problem was simply that the market – and in this case the world market – did not find these profitable. Maybe it is more correct to say that the companies on the world market found others ways of producing the goods by which they could achieve better profits making it impossible to realise the ecologically better ways of production that were invented for a market desiring environmentally friendly goods.

A Plural Economy

In the book *A Human Nature*, Nielsen and Nielsen analyse this challenge between independent systemic processes of production in the modern economy and the necessary natural basis for these processes. They call such an economy, which goes beyond systemic compulsion of accumulation, *a plural economy*. The authors make use of the process of preserving a river valley (named Halkær Ådal) as an example of such a plural economic process.

What is remarkable in their approach is not only that they consider the natural basis to be a precondition for the systemic processes to take place and which also result in processes which destroy the natural basis. They also claim that a long list of social and cultural processes, which are not controlled by the systemic accumulation economy, must be present or available for the latter to take place. Thus it is claimed that a number of social, cultural and ecological processes in relation to society and to nature must be present if the independent systemic accumulative economy is to be possible. Included in relation to these social and cultural processes are the ethical and aesthetic rationalities defined above.

Furthermore, Nielsen and Nielsen claim that the free scope for the systemic economic accumulative processes is narrowed in time along with their improving efficiency. In this way Nielsen and Nielsen point at the possibility of establishing 'spaces' in society that are determined by a different 'logic' than the one of accumulative economy and is thus consequently not destructive in relation to the natural basis.

If such 'spaces' can be viewed as a consequence of the fact that the systemic accumulative economy 'peels off layers' of the economic and the social that cannot be made effective enough to satisfy the market, then it can be claimed that this, in a way, counters what Habermas calls the colonisation of the lifeworld. As a result, it contributes to an expansion of the social, cultural and natural 'space' that *can* escape the systemic accumulative compulsion in the economy and its destructive impact.

Decisively in relation to a discussion on sustainability Nielsen and Nielsen do not ignore the contradictions and destructive logics appearing in outdifferentiations in the capitalist modernisation. They try to include them in their reflections on where and how sustainable processes can be possible. In consequence, their reflections, of course, also include how such processes can be promoted through democratic and participatory political processes.

Furthermore, a vision could be that an establishment of such 'spaces' for sustainability taken together with systemic processes can increase and expand the process of narrowing the space in which the systemic processes take place.

Thus it is also essential that sustainability in such a context can be seen as a partial strategy. Not only in the sense of the gradual creation of sustainable processes, but also partially in relation to those social, cultural and natural spaces in which the strategy will be practicable. Furthermore, it could be partial in the sense of certain decision-making processes, for example those that deal with questions where specific ethical considerations must be taken or certain localities or sites where specific ethical and aesthetic considerations must be included. It may also concern specific populations, their culture and way of living; to which reasoned arguments against systemic invasion can be made. Politically such a framework can be imagined as exemplary initial stages towards a plural economy for or in society.

A New Agenda for Sustainability on Global and Local Levels

The criteria for a new agenda for sustainability must be that the systemic rationalities as well as those rationalities from the lifeworld are taken together and made the basis for actions which include systemic connections, primarily in the economy and the political system.

We must open up the possibility for other types of ethics than utilitarianism. We must open up the possibility for other rationalities than systemic rationalities, and we must open up the possibility for other orientations of actions than goal orientation.

If we by doing so destroy or damage nature it is the same criteria that must be taken into account in deciding to which degree we will accept that destruction and what we will not. It results in only one requirement: thinking efficiency, rightness and truthfulness in a combined way in any action. This can only be done by focusing on the conditions for such reintegration and not by ignoring the reality that modern society separates them.

The point of departure for sustainability must be taken in society's own structure and not in the relation between society and nature. It must not be taken in what nature can cope with, but in society's 'balance' with itself – we could call it the internal dynamic of society – a so-called endurance standard will always be exceeded or broken.

That this is to go beyond the existing capitalist rationality can easily be realised. In such a strategy public insight is a central concept. A first step could be to introduce the same public insight in the actions of business companies as it is practised in public institutions and decision making.

Besides, from what I have said about modernity's outdifferentiation of independent systems, I see *three strategic possibilities* for a new agenda for sustainability.

The *first strategy* is a systemic taking over of what is left of the lifeworld and making it 'rational'. It can be characterised as *free play for the market*. The role of policy in this model is to make the market efficient and prevent market failures. For example, to make the market transparent, remove barriers of different kinds, for instance a refractory 'indigenous population' or other types of resistance. Furthermore, policy can establish the market for tradable pollution permits. It is totally left to the market and the political system to manage the reproduction of society and the reproduction of the natural basis for it. This strategy can presumably be realised, but present scenarios on, for example, the climate catastrophe and its impact is an alarming prospect of the enormous human and economic costs that follow from a reproduction of society based on a nature in such a scenario. It must be an absolutely unacceptable strategy.

The *second strategy* involves political systems taking over the market, which is gradually limited as an independent system. The premises for the economic system are determined by a global political system. Thus it does not necessarily result in an elimination of capitalist accumulation, but its scope is determined politically. This strategy can be named *the political system's dominance over the market system*. This strategy may be realised in the long term. However, in a world consisting of religious and secular states and nations, which can be traditional or modern, plus all possible miscellaneous political systems, this strategy must belong to the category of 'difficult tasks'.

The *third strategy* is the creation of 'spaces' in which the separation or one-sidedness of system and lifeworld is abolished and in which the contradictions between systems and lifeworld are consciously worked on and result in new forms of living. Such 'spaces' can be either specific areas or fields in which the market is set apart, for example nature parks. It can also be specific types of decision making in which influence from the lifeworld is made possible, on the procedure as well as on the result. For example in environmental impact assessments which are understood as a matter of abolishing the above-mentioned contradictions. This strategy can be named *systemic negation*. Globally, examples could be territories, regions or parts of the world in which capitalist modernisation of the social and productive processes do not occur or are not allowed. Rules for such areas, etc. can be determined by the world society, for example the UN, and with participation of the local population.

References

Elling, B. 2008. *Rationality and the Environment: Decision Making in Environmental Politics and Assessment.* London: Earthscan.

Giddens, A. 1990. *The Consequences of Modernity.* Stanford, CA: Stanford University Press.

Habermas, J. 1984. *The Theory of Communicative Action, Vol. 1: Reason and Rationalization of Society.* Cambridge: Polity Press.

Hajer, M. 1996. Ecological Modernisation as Cultural Politics, in *Risk, Environment and Modernity: Towards a New Ecology*, edited by S. Lash, B. Szerszynski and B. Wynne. London; Thousand Oaks; New Delhi: Sage Publications, 246–68.

Honneth, A. 1995. *The Struggle for Recognition: The Moral Grammar of Social Conflicts.* Cambridge: Polity Press.

Macnaghten, P. and Urry, J. 1998. *Contested Natures.* London; Thousand Oaks; New Delhi: Sage Publications.

Mol, A. 1995. *The Refinement of Production: Ecological Modernisation Theory and the Chemical Industry.* Den Haag: CIP-DATA Koninklijke Bibliotheek.

Nielsen, B.S. and Nielsen, K.A. 2006. *En Menneskelig Natur. Aktionsforskning for Bæredygtighed og Politisk Kultur.* Copenhagen: Frydenlund.

Søndergaard, B., Hansen, O.E. and Holm, J. 2003. Ecological Modernisation and Institutional Transformations in the Danish Textile Industry. *Journal of Cleaner Production*, 12(4), 337–52.

Chapter 3

Sustainability as a Tug of War Between Ecological Optimisation and Social Conflict Solution

Jesper Brandt

Introduction

A more sound ecological use of our landscapes has had a central role in landscape and physical planning since the environmental debate started in the 1970s. New dimensions have been added since the Brundtland Report announced the necessity of finding ways of ensuring sustainable development and emphasised changes in social institutions as an integrated part of sustainable development, since social institutions are the most important carriers of both obstacles and enabling factors for the transition process (World Commission on Environment and Development 1987).

From a narrow sociological point of view, social sustainability might be perceived as a system of social regulation mechanisms able to sustain the social coherence of a society or land use system. However, seen in a broader environmental perspective of 'sustainable development', the involvement of social institutions in the implementation of such a development can also be regarded as a question of putting ecology – care of the land – on the agenda for the structural development of the social institutions. They should have a basic interest in the identification and management of thresholds such as different kinds of carrying capacities in complex used cultural landscapes. Built-in structural mechanisms are expected to be the best way of keeping an institution's strategy on the right ecological track.

Many historical studies have been made during the last few decades to explore how past societies have dealt with problems of sustainability. A number of them have focused on the reasons behind the collapses of past civilisations (see, e.g. Tainter 1988) whereas during more recent years the trend has been the opposite: How do we explain that we are still here? Many societies seem to have an incredible ability to survive even long and severe crises (see, e.g. Diamond 2005) and a consensus seems to be developing around the concept of resilience as a framework for the study of the ability of a system to absorb disturbances without changing its structure (Gunderson and Holling 2002).

Since sustainable development is very often connected to the idea of a carrying capacity necessary to avoid environmental problems, historical studies of systems tied to the management of such a carrying capacity are of special interest. In the following, the description of a traditional Faeroese sheep grazing system will be used for a discussion on the relation between ecological sustainability and the regulation of related social conflicts.

The Faeroese village and the related regulation system can be seen as one of the few remnants of an agricultural infield-outfield system which was once the dominant land use system all over Europe up to modern times. A clear concept of sustainability has been a built-in characteristic of this system, which makes a study of its functionality relevant also in a modern context.

The Traditional Faeroese Land Use System

The Faeroe Islands are located in the North Atlantic, between Shetland and Iceland. Their origin is tertiary volcanic, consisting of several kilometres of thick almost horizontal layers of flood basalt lavas with intervening layers of ashes, later formed by ice and strong coastal erosion. This has given rise to a characteristic step-wise relief with marked glacial valleys and steep sea cliffs at exposed coastal positions. The total area is only 1,400 km^2, with a population of about 48,000 inhabitants, living on 17 islands. The largest, Streymoy, also comprises the capital, Tórshavn, where about one-third of the total population lives. Despite the smallness of the Islands, there are in fact about 90 small villages, in some areas amalgamated into larger settlements, along the coasts. Today, almost all of them are connected by an extensive infrastructural system of roads, tunnels, and ferry and helicopter routes. However, until the Second World War, the villages were rather isolated, due to the mountainous topography and difficult climatic and oceanographic conditions, making sea transport a necessary, but also risky way of communication due to very unstable weather conditions and strong tidal currents that exist between the Islands.

Today, the Faeroe Islands are a rich fishery nation, where agriculture contributes less than one per cent of the national product. But until the middle of the nineteenth century, agriculture was the dominant occupation. Although products from milking cows had a decisive role for the local nutrition, sheep breeding held a unique position as it produced an essential product for export and formed the main basis of taxation on the Islands, which for a long period was subject to different trade monopolies and substantial Crown estate. Since the increase in wool prices from the fourteenth century onwards, strong inducements have risen to increase the exploitation of the acreage available to the Faeroese communities, primarily by means of increased sheep breeding. Wool and its products thus amounted to approximately 90 per cent of exports during the entire eighteenth century.

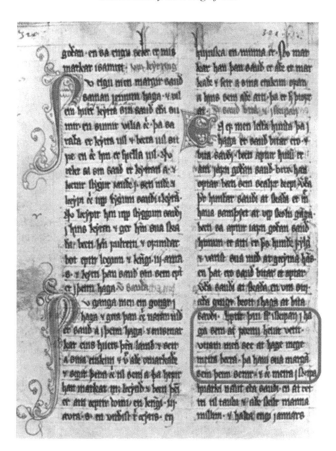

Figure 3.1   The Sheep Letter, 1298

Note: Version Lund 15, 133v (Matras, Poulsen et al. 1971), indicating that 'the number of sheep to be kept on an area of pasture land shall remain the same as it was in previous times. If they agree that it can accommodate more, then they can have as many (sheep) as they can agree upon, and every man can have as many sheep, as his share of the property can justify'.

The Sheep Letter

As early as 1298, a special law for the Faeroe Islands, Seyðabrevit (the Sheep Letter) was passed, which, among other things, stated that 'the number of sheep to be kept on an area of pasture land shall remain the same as it was in previous times' (Seydabrævid 1971: 49, see Figure 3.1). This number, in the Faeroese language called *skipan*, which may be historically related to the English word 'shipping', expressed the carrying capacity of each individual location, and to this day it is used as an expression of the optimum carrying capacity for the various parts of the Islands. Additional skipans for cows, horses, dogs and geese were developed as well.

Each village existed as a typical infield-outfield agricultural system with a little infield located by the sea and surrounded by a one-metre high stone wall dividing the infield from the surrounding extensive outfields.

The most important farming activity was cattle and sheep rearing. There was a close relationship between these two stocks and the use of infield and outfield; in the summer the cows grazed the lower outfield, while the sheep were shepherded to the upper outfield. In winter, the cows were confined to the stable and fed on hay harvested the previous summer in the infield, while the sheep grazed the lower part of the outfield and some of them also the fields of the infield. There were no fences around the extensive outfields before the 1960s. Nevertheless the grazing depended on a very distinct but flexible territorial structure of pastures, each carrying between 12 and 90 ewes plus their lambs, living in flocks showing a very distinct territorial behaviour.

The Sheep Grazing System of the Outfields

Not only were the carrying capacities of the approximately 250 outfields of the Islands carefully stated and used for taxation, but also at the local level, within the single outfield, a detailed regulation of the number of sheep in each flock was maintained. As an example, the structure of the grazing around the mountain Stóra Fjallið, between the villages of Húsavík and Dalur at Sandoy, is shown in Figure 3.2. Here, about 500 ewes grazed in a very flexible manner, even around the outfield border between the villages. In a trial which took place in 1753 it was stated that:

> In Christian love it should be tolerated when sheep from Dalur are drifting into the outfield of Húsavík, and they should be left there freely, until they can escape again without any hazard. Also men from Dalur should accept, when sheep from Húsavík against their will are drifting into the outfield of Dalur, and both parts should take their sheep back in their own outfield at the first opportunity (Landbokommissionen 1911).

Obviously the concept of skipan could be regarded as a measure to regulate sustainability in the traditional Faeroese infield-outfield system, based on an intended ecological optimisation process behind the structure. This optimisation could partly be seen as a question of finding a balance between, on the one hand, land use and productivity of the infield and outfield and, on the other, the need for risk-minimisation as regards the availability of fodder for cows and sheep. Due to the agriculturally marginal position of the Faeroe Islands and the very limited amount of well-drained lowlands, the production of fodder was reserved exclusively for the cows, leaving sheep breeding as a more or less risky pursuit. The oceanic

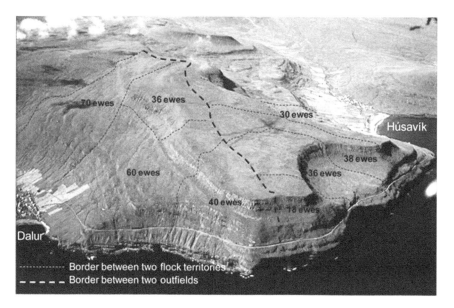

Figure 3.2   Sheep grazing around Stóra Fjall between the villages Húsavík and Dalur at Sandoy, Faeroe Islands

Source: Jesper Brandt.

position gives an average July temperature of only 11°C, but with mild winters of an average of 4°C in January, outdoor grazing was usually possible all year round. However, in some years a severe winter occurs in which frost and snow affect the pastures for a longer period. This could have disastrous consequences, especially after a summer period with modest vegetation growth. If more than one-third of the sheep population died as a result, it was called a *felli* (plural: *fellir*). According to a historical source from 1783, such a felli was expected to occur on average once every 14 years (Svabo 1959: 743).

The occurrence of felli certainly influenced the optimisation process that had to include this risk, since it took years before the size of the stock returned to the normal level.

The principle of risk minimisation is explained in Figure 3.3: On average only one lamb per ewe could be expected each year, giving a slaughter percentage of around 50 per cent. Thus, with a stock of 200 ewes producing 200 lambs per year in an outfield, one could expect to slaughter 200 sheep per year. For a period of 28 years as shown in Figure 3.3, it should potentially give a possible slaughter of, all in all, 5,600 sheep. But with an average of two felli in this period, the slaughter would be reduced considerably, in this scenario by 11 per cent, to 4,976. One might raise the skipan, e.g. to 225 ewes, giving a potential slaughter of 6,300, but due to the more dense grazing, the consequences of a felli would be much more severe and prolonged, here resulting in an overall reduction of the slaughter by

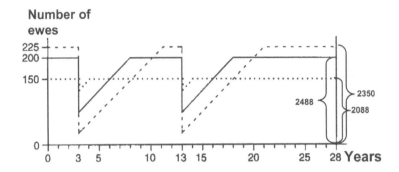

Figure 3.3   The principle of optimisation of the sheep carrying capacity ('skipan') in a time perspective

25 per cent (to 4,700) less than before. One might alternatively lower the skipan, e.g. to 150, thereby reducing the loss during a felli considerably, but also in this case one would end up with a lower result, showing that a skipan of around 200 represents a historically developed optimum carrying capacity.

These principles of optimisation had to be adapted to the local landscape conditions. This adaptation has been studied through a regional investigation of a larger area. Figure 3.4 shows the layout of the 87 pastures in the 15 outfields of the eastern part of Sandoy Island. The pastures around Stóra Fjallið are encircled. This information was gathered at the end of the 1970s through interviews with old shepherds having been active since the First World War. Although slightly modified, the pastures have, in principle, been rather constant for at least 500 years. The figures indicate the skipan of each pasture. In general, each flock should stick to its pasture the year around. Some flocks had, however, the right to graze the infield in wintertime and their pastures are indicated with a raster on the map.

The adaptation process was studied quantitatively through an 'up-side-down' landscape ecological analysis of the relation between vegetation productivity and carrying capacity: often an estimation of the carrying capacity in a grassland system is estimated by measuring the productivity (e.g. as digestive organic matter (DOM)) of the different vegetation types under different geo-ecological conditions, which are summed up based on a detailed vegetation survey. But instead of measuring the productivity of the vegetation directly, e.g. through the cutting of vegetation samples, the known historical data (gathered from taxation records or estimations given by old shepherds) of skipan were used to estimate the productivity of the vegetation. An optimisation modelling of the relation between the productivity of the sheep grazing system (based on the number of slaughtered sheep) and the basic vegetation production of the landscape was developed by using least-square method and linear programming. This permitted not only the estimation of the efficiency of the adaptation to the local

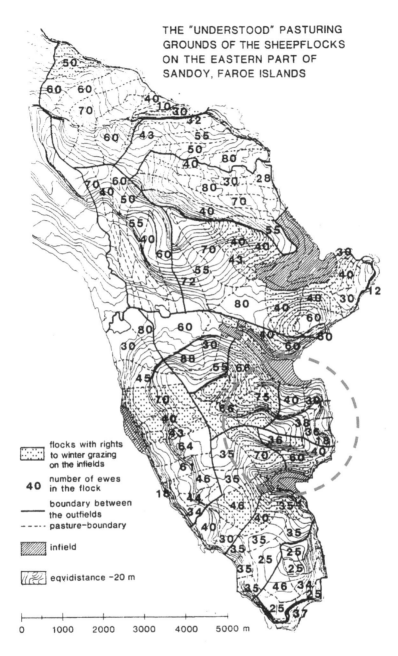

Figure 3.4 **The pasturing grounds on the eastern part of Sandoy, Faeroe Island, including indication of their carrying capacity for sheep ('skipan')**

Source: Survey based on interview with local shepherds in the late 1970s (Brandt 1984).

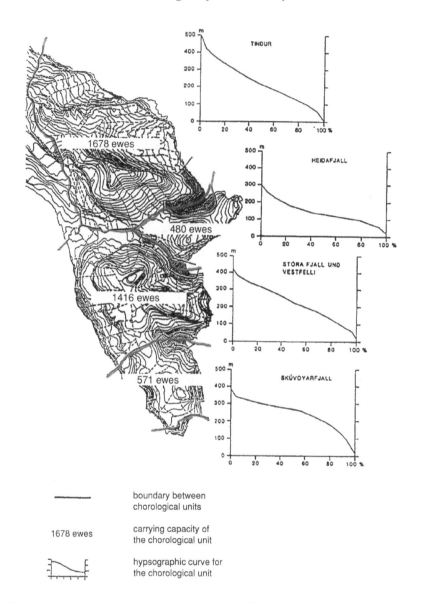

Figure 3.5   A layout of the basic principle of adaption to the local landscape

landscape conditions, but also to localise and evaluate important bottlenecks in the grazing system. For further details on the method see (Brandt 1984, Brandt 1992).

The basic principle of adaptation to the local landscape is outlined in Figure 3.5, showing how four main landscape units of the eastern part of Sandoy can be

Figure 3.6   The most northern outfield of the eastern part of Sandoy

distinguished, each of them showing a characteristic landscape chorology, decisive for the sheep grazing through its ability to deliver suitable pasture grounds for all seasons and all weather conditions. Each flock should preferably have different vegetation types at different heights at its disposal, which can be expressed through hypsographic curves that synthesise the local climatic and vegetation conditions to which the layout of pasturing grounds have been adapted.

A closer look at the two most contrasting landscape units can clarify the importance of the adaptation process.

In the northern part of the area around the peak of Tindur, see Figure 3.6, the composition of the landscape allows for a distribution of different grassland types giving bearable conditions under most weather conditions, and tolerable winter grazing during harsh weather conditions. The layout of the pastures has clearly been built up in strips running from the well-drained grass slopes good for summer grazing near the summits down to the harsh heather and moor vegetation near the sea or lowland areas of, e.g. the rush *Juncus Squarrosus*, which are not so attractive for the sheep, but on the other hand safe in most winter situations.

In the southern part around Skúvoyarfjall, see Figure 3.7, the topographical conditions are less favourable, since the heather and moor vegetation, constituting a safe winter grazing resource, are situated on the high plateau, and the luxuriant grass slopes, most suitable for summer grazing, are located down along the coast. Situations with felli in winter and early spring when alternating frost and thaw

Figure 3.7   The most southern outfield of the eastern part of Sandoy

might spoil the grass of the exposed slopes has obviously given rise to a more complicated pattern of pasture inhabited by rather small flocks which are easier to manage under these difficult conditions.

In general the system worked, but a felli in this area could be disastrous since no lowland winter pastures usable under harsh conditions were available. Therefore the skipan had to be kept at a rather low level.

Through the modelling of the adaptation of the territorial system to the landscape it could be estimated that the grazing system in the southern area was able to use only half of the outfields' mean annual production of DOM, whereas this percentage was raised to two-thirds in the northern outfield, thus showing a more straightforward adaptation of the grazing system to the landscape composition.

The Sheep Letter as Social Regulation

The foregoing analysis is founded on the basic assumption that social mechanisms of optimal and sustainable use of the grazing potential were built-in characteristics of the Faeroe sheep grazing system, legally grounded in the Sheep Letter of 1298. Serious reasons to question this assumption exist. Both an analysis of the development in legislation as well as historical studies of the development of skipan indicate that a stable ecological sustainability is a too simple explanation for the functionality of the skipan.

In outline, the regulation of the Sheep Letter proved so stable over the centuries that it has been taken as a proof of an origin of the regulation system much older than thirteenth century (Bærentsen 1911). But, nevertheless, the Sheep Letter was renewed several times, especially concerning the ownership of sheep. Two main types of ownership have existed in parallel throughout history. Although an outfield with several owners was always a common, the sheep of the common could be individually owned in number according to the owner's share of the Marketal – this was called *kenning*. The sheep could also be collectively owned, and the ownership then expressed in the share of the total slaughter (including wool) equivalent to the owner's share of the Marketal – this was called *felag*. At the time of the Sheep Letter, kenning was absolutely dominant, which obviously influenced both language and rule-setting. But this changed, especially up to the eighteenth century, when the growing fragmentation of private property, obviously, in some areas, made it difficult to manage the grazing system. Smallholders could invest time in the isolated herding of their own sheep, which was in disfavour especially with the King's tenants, often clergymen and other official authorities, having the right to use about half of the total land of the Faeroes (Landbokommissionen 1911: addition, 76). Therefore kenning was forbidden in 1659 and felag forced through with the revisions of the Sheep Letter.

In connection with a revision in 1757, a report from the county mayors was produced to support the future legal administration. Figure 3.8 is based on this report. It shows that although felag was dominant, there was still a widespread use of kenning, especially in the more remote parts of the Islands. Not only were these under less influence of the central authorities, but these areas (like the southernmost Island Sud017oy) were also historically characterised by a larger share of the land being owned by smallholders.

The trend towards felag changed again during the nineteenth century. Expansion of the infield following population growth from the beginning of the nineteenth century (at least partly related to an emergence of a growing class of smallholders and landless peoples involved in fishery) was difficult to handle due to the common property rights of the outfields. High prices on woollen products made it in principle possible for this new class to also make a living from knitting socks and sweaters. To maintain the labour supply for the bigger farms, legislation made it illegal to collect the necessary wool in the outfields, although it was often dropped by the sheep. Additional legislation was passed, forbidding young people to marry without having a legal occupation; knitting was not such an occupation! But in the end, the growing interests in keeping low reproduction costs for smallholders partly engaged in fishery took over and kenning, most suited for the smallholders, was legalised again in 1866.

The development of the legislation around felag shows that social affairs and conflicts had a considerable influence on the management of the skipan.

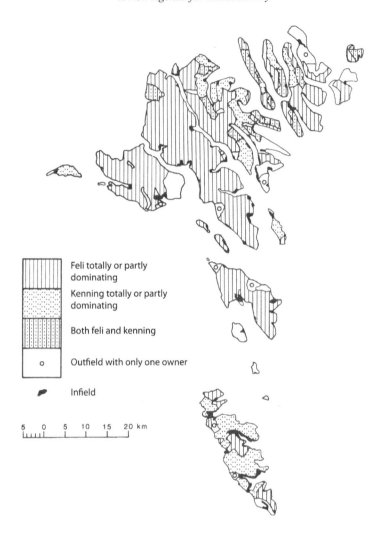

**Figure 3.8   The distribution of separate ownership ('kenning') and common
             ownership ('feli') at the Faeroe Islands in 1758**

Source: Reports from the regional governors. Føroys Landsskjalasavn, Tórshavn.

It could, of course, also be interpreted, ecologically, as a measure to ensure an
optimal resource use, since joint management was based on the election of joint
shepherds who had total control over the pasture, ensuring the necessary measures
to sustain the productivity. Nobody, not even the owners, was allowed to cross the
outfield without the permission of the shepherd. So, the shepherds were extremely
important people in the village. In the past there was normally one shepherd per
outfield, of which there could be several in a village. He was elected by the owners,

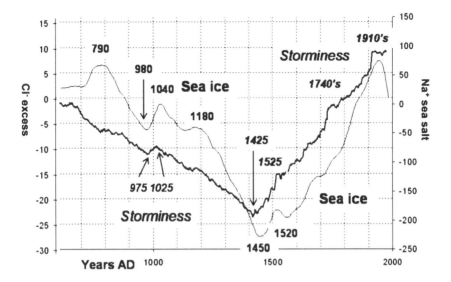

Figure 3.9   Climate diagram

Note: Cumulative records of annual deviation from the longterm mean of the time series for the proxy records of Greenland Sea/Davis Strait sea ice extent and North Atlantic storminess. The main interpretation of the curves are described by Dugmore, Borthwick et al. (2007) in the following way: 'Crucially, a shift in the direction of change can be used to mark the point at which prior change fails to predict future change and the gradient of the graph can be used to indicate a cumulative rate of change. Cultural stresses enhanced by climate (with potential impacts on landscape and settlement) may therefore be most likely where the gradient of the graph is steep, or where the slope changes direction'.

Source: Data from Greenland Ice Sheet Project 2. http://www.gisp2.sr.unh.edu/ (Dugmore, Borthwick et al. 2007).

according to their share of the total number of marks. Thus, an owner or King's tenant having more than half the total value of an outfield (measured in mark=16 gylden=320 skind) could decide who should be the shepherds.

The reporting of widespread overgrazing and erosion as well as statistics on the development of skipan seems instantly to contradict such an interpretation.

Throughout history there are many reports on the lowering of carrying capacity and widespread soil erosion (Debes 1673: 117, Svabo 1959) (Landbokommissionen 1911: 69). But sheep and erosion have not necessarily been related to each other. Sheep have been such an integrated part of the Faeroese cultural landscape since the Celtic settlement in the sixth century, so it must have been difficult to judge how the geo-ecology would have developed without sheep: There have always been signs of erosion on the Faeroe Islands as well as sheep all over. Sheep have been so much a part of the cultural landscape that they have been considered a part of nature.

Table 3.1   The development of the total sheep flock on the Faeroe Islands 1600–1988

App. year	1600		1780		1870		1988	
		(Index)		(Index)		(Index)		(Index)
Suðuroy	17,578	(100)	15,600	(89)	11,949	(68)	10,111	(58)
Sandoy	12,412	(100)	10,375	(84)	7,760	(63)	7,738	(62)
Vágar	11,220	(100)	8,820	(79)	6,730	(60)	7,540	(67)
Streymoy	21,740	(100)	16,110	(74)	15,549	(72)	14,577	(67)
Eysturoy	19,840	(100)	13,703	(69)	13,824	(70)	14,155	(71)
Nordoy	13,759	(100)	10,931	(79)	8,296	(60)	10,026	(73)
Total	96,549	(100)	75,539	(78)	64,108	(66)	64,147	(66)

Note: The numbers are all based on the known skipan, not statistical counting. This is in reality also the case for 1988, although it should have been based on a counting.
Sources: (Brandt 1987) and Árbók fyri Føroyar 1991.

It is, however, a hard fact that, on average, skipan was reduced by one-third from the beginning of the seventeenth century to the end of the nineteenth century, although at different rates in the different parts of the Islands (see Table 3.1). The stabilisation within the last century probably implies continued overgrazing due to the introduction of modern medicine, winter shelters and a growing amount of imported winter fodder. Productivity has, in general, increased in terms of number of slaughtered sheep and an increase in slaughter weight. Decreases in the active sheepherding and maintenance of the outfields (especially drainage) paired with the use of winter fodder might have increased the pressure on low-laying pastures, and sub-optimal use of more remote areas. Such overgrazed 'feeding lots' are well-known in grazing systems.

Apparently the Sheep Letter only partly regulated an ecological carrying capacity and was primarily concerned with social conflicts at the local level of a single island due to uneven grazing pressure. The territorial system of grazing was upheld by the principle of the different flocks being deliberately 'shepherded against each other', that is, the grazing pressure of each flock had to be kept at the same level so that there was no reason for a systematic trespassing from one ground to another.

However, if the owners were able to agree on an over-exploitation of all the pastures, they were allowed to do it. In the short term this might have been seen as a sign of growing productivity, resulting in higher taxation revenue. Because the quotation from the Sheep Letter (see Figure 3.1), continues as follows: 'If the owners agree that it can accommodate more, then they can have as many (sheep) as they can agree upon, and every man can have as many sheep, as his share of the property can justify'.

This indicates that the Sheep Letter primarily was established to solve social conflict between land owners.

Emphasis on a social interpretation of the use of skipan is also supported by the fact that considerable differences in the weight and slaughter percentage exist between the different islands, thus indicating very different degrees of grazing pressure. These differences can at least partly also be linked to obvious variations in soil erosion between the different islands.

Skipan as Ecological Wisdom

Although the Sheep Letter has primarily to be seen as a form of social regulation, it does not necessarily mean that the concept of skipan did not express the endeavour of the local land use system to ensure a sustainable use of the natural resources.

The skipan was not just passed down through the centuries. In fact, marked departures from the general trend of a falling skipan did arise. In a very interesting report from 1783, a young student Johan Christian Svabo (1783) commented on the general decline of the skipan since the beginning of the seventeenth century, but added a handful of exceptions most notably the Island of Nolsoy, by the entrance to Tórshavn, 'where the old skipan was 900, but due to the skill, industry and good management of the inhabitants had been raised to 1,100' (Svabo 1959: 771). Ninety years later, however, the skipan of Nolsoy was registered at 920 (Taxation 1873).

Concrete measures to keep the skipan were also a known practice. In some villages it was a standard procedure to make an evaluation of the tallow weight of the sheep by the first collection of sheep in the autumn. If it was below a certain level, the number of sheep to be slaughtered by the second collection was increased to lower the number of sheep grazing the obviously less productive area (personal communication, Rolf Guttesen).

Thus, an active ecological wisdom concerning sustainability did exist, and was a part of the management practice of the shepherd. He was, of course, dependent on the owner, and it is conceivable that his experience could have difficulties in being acknowledged. But his knowledge was not just attached to the outfield he managed. He cooperated closely with all the other shepherds for many practical reasons, first of all to prevent all the troubles that could turn up if the grazing pressure of the different parts of an island was dissimilar. His education was closely related to this cooperation thus giving the shepherds a powerful voice provided they were able to cooperate. The historically developed ecological wisdom of the grazing system was born out of the cooperation of these shepherds. However, their experience could only be based on the past. If trends in nature changed, they might fail.

The marked differences in the development of skipan at the regional and local levels between the sixteenth and nineteenth century might have been related to more basic ecological problems in the form of increased difficulties in risk assessment due to climatic changes.

Rather good documentation of the little ice age, approximately 1300–1870, as well as the preceding medieval warm period, approximately 800–1300, exists. It seems reasonable in some way to relate this development to the development in skipan.

In an historical assessment of climate impacts on settlement, it has been suggested that many complex societies have a marked resilience to droughts and other occasional events within a time span of some decades, but may also exhibit marked responses to events at a level of many decades to centuries (de Menocal 2001).

In a North Atlantic context (Dugmore, Borthwick et al. 2007) have drawn the attention to the necessary strategy behind this resilience, arguing that 'the ability of human systems to accommodate or adapt to bad seasons may be primarily constrained by their predictability on the decadal scale'. They especially focus on the identification of the validity of past experience concerning risk-benefit assessments of hazardous undertakings since the assessment of predictability might be just as critical for a land use system as the mere ability to adapt to periods of unfavourable conditions. They have used the cumulative deviations from the means of the Greenland ice core storm frequency proxy (GISP2 Na+) and sea ice proxy (GISP2 chloride excess) to identify episodes of unpredictable change in the storminess of the North Atlantic within the last 1,400 years, see Figure 3.9.

Most remarkable is a clear shift after 1425, but the authors also emphasise the fact that 'the rapid changes of the mid-eighteenth century coincide with one of the extreme periods of the last 1400 years'. They stress that 'In terms of predictability, another key climatic change is apparent in the storminess record after 1425 AD. Since the early fifteenth century the record has become more changeable with more extreme years'.

The relevance of these data for an evaluation of changes in the climatic conditions for the Faeroese sheep grazing, and especially for an assessment of possible changes in the occurrence of fellir as an important part of the determination of the skipan, can certainly be questioned: The length of the winter season and the productivity of the foregoing summer might be more important than storminess.

But their might very well be a relation between this storminess and Svabo's reference to a 'felli' on average every 14 years that was closely related to the long-termed valuation of the carrying capacity in the form of skipan. He describes it as a risk assessment, calculated by the tradition of 'the old men', adding that a calculation from the last 100 years confirms this assessment to be rather close. But he also adds that such a calculation cannot be made precisely. The little ice age might have passed outside the horizon of an experience-based hazard-inclusive assessment of the carrying capacity of the outfields for sheep breeding. Obviously, a late medieval shift in overall agricultural strategy took place on the Faeroes, replacing a former rather intensive dairy system, including shielings, or summer farms, using widespread tillage in the (later) outfields (Brandt and Guttesen 1981), with the later system of more contracted infield activities combined with extensive and exclusive sheep grazing in the outfield. Despite the apparent coincidence of this shift and the change in climate observed by Dugmore, Borthwick et al. (2007), the shift seems rather to be related to changing trade relations and growing wool prices, accentuating sheep farming at the expense of other types of agricultural activities. A less predictable impact of the shift might, however, have been related to the changes in the circulation of nutrient. Adderley and Simpson (2005) argue that yields of cereals and fodder were optimised by the manuring strategies on the Faeroes around 1200 AD and Dugmore, Borthwick et al. (2007) suggest that the later changes in the climate record should be visible in the landscape of the later system due to the cessation of the widespread use of manure to compensate for yield loss.

Thus, one can imagine that the rapid climatic 'changes of the mid-eighteenth century' (Dugmore, Borthwick et al. 2007) might have represented a serious challenge to the authority of risk assessment based on former experiences, giving a defence for more short-termed interpretations of skipan related merely to social conflicts.

Conclusion

Diverging facts can give rise to different interpretations of the historical Faeroese sustainability concept of skipan. Basically, it has regulated a long-termed optimal use of the grazing potential of the Islands under shifting economic conditions. The variable weather conditions have made risk assessment of the grazing system through the judgement and constant evaluation of a long-termed sustainable level of skipan a crucial part of the regulation of the land use system. The shepherds of an island formed the expertise, developing a variety of indicators of grazing pressure and its relation to short-termed and long-termed regulations of the sheep stock at a local and regional level. Their cooperative effort based on a common experienced process of environmental learning was of utmost importance, especially for the minimisation of the consequences of extreme winter and spring situations, the so-called *fellir*. However, widespread soil erosion in combination with a notable decrease in the carrying capacity for the main grazing animal, sheep, during the last 400 years makes the historical efficiency of this way of sustainability regulation questionable. Also, the focus of the Sheep Letter (1298) on social conflict solution, admitting the right to raise the stock unlimited, if the owners could agree on it, makes the ecological interpretation of skipan less inevitable.

The long-termed changes in the climate since medieval time must have had a profound influence on the experience-based possibilities of risk assessment and optimal resource use based on a judgement of the local optimal level of skipan. This might have challenged the authority of the institution of the co-operating shepherds for an environmentally/ecologically-based regulation of skipan compared to the short-sighted advantages of a regulation strategy based solely on social conflict solution among the land owners. The land owners could be active shepherds too but, in general, and probably especially among the many officials being King's tenants on relatively big holdings this was not the case. More historical research in the socialisation and cooperation among shepherds as well as in the social relation between shepherds and land owners is, however, needed to shed light on the historical weighing of ecological and social sustainability. But if the thesis of a shift towards a primarily social regulation of the skipan due to failing risk assessments eventually following climate changes is right, it will form a good example of 'The Pathology of Regional Resource and Ecosystem Management' described by Gunderson and Holling stating that 'mediation among stakeholders is irrelevant if it is based on ignorance of the integrated character of nature and people' (Gunderson and Holling 2002: 8).

References

Adderley, P.W. and Simpson, I.A. 2005. Early-norse Home-field Productivity in the Faeroe Islands. *Human Ecology*, 33(5), 711–36.

Brandt, J. 1984. *Landscape Ecological Information through Statistical Analysis of the Territorial Structure of a Sheep Grazing System, Faroe Islands*. Proceedings of the first International Seminar of the International Association for Landscape Ecology (IALE), Vol III. Roskilde: Roskilde University, 15–19 October 1984.

Brandt, J. 1992. Schafzucht auf den Färöern – zur Gewinnung landschaftsökologischer Informationen durch statistische Analyse eines Landnutzungssystems. *Petermanns Geographische Mitteilungen*, 136(5+6), 235–49.

Brandt, J. and R. Guttesen 1981. *Changes of the Rural Landscape on the Faroe Islands in the Middle Ages*. Proceedings: Collected papers presented at the Permanent European Conference for the Study of the Rural Landscape held at Roskilde, Denmark 3–9 June 1979, Copenhagen.

Bærentsen, C. 1911, Ejendomsforholdene i Bygden Sand, in *Forslag og Betænkninger Afgivne af den Færøske Landbokommission Nedsat i Henhold til Lov af 13 Marts 1908. Tillæg til Forslag og Betænkninger.* Copenhagen: Schultz, 463–581.

Debes, L. 1673/1965. *Færoæ & Færoa Reserata*. Kjøbenhafn/Thorshavn: Einars Prent og Forlag (Reprint).

de Menocal, P.B. 2001. Cultural Responses to Climate Change During the Late Holocene. *Science*, 292(5517), 667–73.

Diamond, J. 2005. *Collapse: How Societies Choose to Fail or Survive*. London: Penguin Books.

Dugmore, A.J., Borthwick, D.M., Church, M.J., Dawson, A., Edwards, K.J., Keller, C., Mayewski, P., McGovern, T.H., Mairs, K.-A. and Sveinbjarnardóttir, G. 2007. The Role of Climate in Settlement and Landscape Change in the North Atlantic Islands: An Assessment of Cumulative Deviations in High-resolution Proxy Climate Records. *Human Ecology*, 35(2), 169–78.

Gunderson, L.H. and Holling, C.S. 2002. *Panarchy: Understanding Transformations in Human and Natural Systems*. Washington, DC: Island Press.

Landbokommissionen. 1911. *Forslag og Betænkninger Afgivne af den Færøske Landbokommission Nedsat i Henhold til Lov af 13 Marts 1908*. Copenhagen: Schultz.

Seydabrævid. 1971. Edited by C. Matras, J.H.W. Poulsen and U. Zachariasen. Tórshavn: Føroya Fródskaparfelag

Svabo, J.C. 1959. *Indberetninger fra en Reise i Færøe 1781 og 1782*. Copenhagen: C.A. Reitzels Boghandel (Reprint: Ed. N. Djurhuus).

Tainter, J. 1988. *The Collapse of Complex Societies*. Cambridge: Cambridge University Press.

Taxation 1873/1973. *Protokol over den i Henhold til Lov Angaaende en ny Skyldsætning_af Jorderne paa Færøerne af 29de Marts 1867 Foretagne Taxation af Bemeldte Jorder*. Thorshavn: Bladstarv (Reprint).

World Commission on Environment and Development. 1987. *Our Common Future*. Oxford: Oxford University Press.

Chapter 4

Sustainability, Biodiversity and the Ethical Aspects of the Deliberate Release of GMOs

Peder Agger and Erling Jelsøe

Introduction

The purpose of this chapter is to reflect upon the role of the ethical aspects of sustainable development. Although this is often immanent because of the normative basis of the concept, the ethical aspect is often largely overlooked. But ethics is a vision or a conception of the good life. We are united in our view of human nature as something that ought to continue as long as possible. 'Without ethics the demand for sustainable development becomes an unfounded claim. And although there often are arguments for a sustainable development without mentioning ethics, it will stay as a tacit precondition, attracting meaning to the argumentation from the beginning to the end' (our translation of Kemp 2000: 14).

The basic ethical ideas behind sustainable development in the Brundtland Report are, according to Kamara and Coff (2006), the following:

- *Meeting the needs:* Present society should take care of the needs of the poor and future generations.
- *Social fairness:* A fair distribution of resources within global populations is important in itself as it is for the development of environmental sustainability.
- *Maintenance of natural resources and nature:* Scarcity of natural resources sets limits. Regenerative capacity should be maintained and biodiversity preserved.
- *Sustainable economy:* Revitalisation of economic growth based on new qualities like fair distribution, producing more with less, and fighting poverty and environmental degradation.

Just as somebody might have thought that the dispute concerning the deliberate release of genetically modified organisms (GMOs) was fading away, there may be a revival or at least a prolonged and stubborn continuation. In the 1980s and 1990s we saw an extraordinary growth in biotechnology. New research findings paved the way for an unprecedented development from which new possibilities emerged. Yet, politicians and managers were unprepared to tackle the related ethical issues. Since then, biotechnical advisory boards have been established in all European countries. Some were staffed with experts, others, like the Danish and the Norwegian ones, include laymen and terms of reference that require both conventional expert advisers and the stimulation of public debate on these new issues (Lindsay et al. 2001).

The new technologies have challenged deep-rooted cultural norms related to human life and nature, a challenge which will probably continue for the years to come. The disputes started with issues that have been debated for several years like GMOs, in vitro fertilisation (IVF), cloning, and stem cell research and therapy. New themes are coming onto the agenda such as those concerning the ethical aspects of the research and implementation of cyborgs and robot technology. And old themes such as issues relating to the sale of meat from cloned animals are no longer discussed; such meat is already sold (Gaskell 2007) and the deliberate release of genetically modified plants (GMPs) on European fields is at present not an issue. But what mistakenly looks like a ceasefire might be a process of increasing tension where GMPs are being introduced in one country after the other (though still at a very low level) while public resistance remains unchanged.

In the following, we will begin by elaborating the interpretation of the concepts of sustainable development and of biodiversity and the interrelationship between these two. Then we will turn to the Danish Council of Ethics,[1] which works, among other things, with ethical problems in the field of biotechnology pertaining to human beings, nature, the environment and foodstuffs. Recently, the Council was asked by the Danish Minister of the Environment to look at the criteria for releasing GMOs. We will reflect upon what might be the drivers behind the Minister's letter, we will also refer to the Council's discussion of the concept of 'utility', and the conclusions the Council came to before we provide some final reflections on what influence these observations may have for praxis as scientists and our everyday life as citizens, and what it might bring to a new sustainability agenda.

The Concept of Sustainable Development

Sustainable development is a concept that has, you might say, an inborn doubleness, i.e. a discrepancy between system stability on the one side and dynamics of nature and society on the other. Sometimes this might be taken as an insurmountable incompatibility between the 'balance of nature' and an unavoidable, further, technical and economic development of society. But it might also be taken as the concept that connects stability and development in the many forms and varieties in which it might exist in space and time.

Sustainable development has always had a special appeal to people that work with nature conservation and management. There are two reasons for this. The

---

1 The Danish Council of Ethics provides advice to the Danish Parliament and raises public debate about ethical problems in the field of biotechnology (www.etiskraad.dk). Peder Agger is the current chair of the council.

first is that the concept forces a broader and more far-sighted framing of the debate than we have been used to. The other is its complexity. If biodiversity should be understood as an element of sustainable development and managed properly, it is important that it is seen as a broad, long-term and complex matter.

This has to do with space and time and our ability to comprehend the fact that the world develops at many different levels and at the same time has many different rhythms, while we as human beings only live and intuitively understand what is going on within our own timescale. As living beings we are adapted to our environment and bound to the scale determined by our own dimensions, i.e. our body size, range of action and lifespan. We can call it the human scale, this has a range from a few tenths of a millimetre to lets say 5,000 to 10,000 km, from tenths of a gram to say 1,000 to 10,000 tons and from a tenth of a second to let us say 1,000 or 10,000 years. All lengths, weights and time spans beyond these limits are not immediately intelligible and remain abstractions for us.

We can, of course, intellectually and instrumentally break these boundaries. We can enter the nano-technological world by Scanning Tunnel Microscopes. We can survey the globe from satellites, and with radio telescopes we can look far out into the universe. Computers can help us to handle and synthesise large amounts of data and thus create order out of chaos. In this way, we can extend our insight into the world and our possibilities and abilities to comprehend and handle it both in the human and 'extra-human' scales. The German biologist Hans Mohr has similarly introduced the concepts of microcosmos and macrocosmos to denote the spheres that are outside the scope of human sensation and cognitive horizon, whereas mesocosmos is the sphere, which is accessible through the everyday experience of time, space, substance and causality (Mohr 1987, see also Coff 2006a). The point is that ordinary people, laymen (and we all are in most aspects of life and society) have to, or are confined to, act within the human scale. And, therefore, we are on the one hand easily left to the experts, and on the other we are tempted by politicians who are, or pretend to be, as 'simple minded' as we ourselves are (take the inertness in the climate debate as an example). This is a problem as many of the changes we induce into our environment are, so to say, active on all scales simultaneously.

If we then add the complexity that is also common for many environmental problems, it should be understandable why many environmental problems are often not properly understood. The human brain is not able to handle much more than four to five different factors at a time (Schutz 1989). But in the real world, many more are often in operation. Differences in scale and complexes including many factors are at play when matters like the reduction in biodiversity are to be comprehended. In everyday politics they are, therefore, seldom given the recognition they deserve – this raises a democratic problem. Here the concept of sustainable development comes in as a means to overcome some of the problems. Sustainable development offers a relevant framing of the debate on broad and far-sighted issues. It is a normative way to organise politics at a higher level, a frame or strategy, a planning of planning that helps us to organise our comprehension of complex systems across scales of time and space.

The Concept of Biodiversity

The concept of biodiversity has a recent history that is parallel to the history of sustainable development. Biodiversity was also on the agenda at the Rio Summit in 1992 where The Convention on Biological Diversity was signed by 155 nations. Biodiversity refers to the variation among living things and can be taken as synonymous with life on earth. The first Danish biodiversity strategy states 'that the goal of our entire nature and environmental protection effort is the preservation of our biodiversity' (Prip et al. 1996). The Convention on Biological Diversity itself defines biodiversity as follows:

> Biological diversity means the variability among living organisms from all sources including, inter alia, terrestrial, marine and other aquatic ecosystems and the ecological complexes of which they are a part; this includes diversity within species, between species and of ecosystems. (Convention on Biological Diversity 1992)

Usually biodiversity is treated at three levels: the genetic level, the species level and at the ecosystem level, although it in principle also exists beyond and in-between, e.g. at the population, community and landscape levels.

This implies that not only wildlife but also domestic species and GMOs are covered by the notion. Further, biodiversity is a relative notion, to be specified with reference to a certain level of organisation and to a certain area. At the genetic level, it may be defined in terms of variation in genotypes or genes. Thus defined, biodiversity may also be applied within a given species, which, for example, is relevant in connection with domestic species where there is a concern to preserve genetic diversity as a resource for future breeds, or anxiety of possible dispersal of genes from GMOs to the populations of wild allied species. So the concept of biodiversity is not a simple one. Three things add considerably to the complexity of the issue. One is the vagueness of the concept of 'a species'. The other is the uncertainty about how many species we have or have lost. And the third is the blurred concept of what is an indigenous species.

Species are in everyday life taken as indisputable and distinct. No one can be in doubt about what a dog or a cat is. But if we go deeper into the taxonomic literature, it becomes increasingly clear that species can only be taken as landmarks in a continuum of life. This is true so far as taxonomists today operate with several conceptions of a species: The conventional biological concept where two individuals belong to the same species if they can produce full fertile offspring (Mayer 1942), i.e. they have a common gene pool that is isolated from others, and the phylogenetic concept where a species is defined as the smallest diagnosable cluster of individual organisms within which there is a parental pattern ancestry and descent. Further, the taxonomists operate with notions of semi-species, subspecies, races, ecotypes, etc. And modern molecular biology has further dissolved the classical concept of species again and again by splitting up or amalgamating what previously were taken as good distinct species. These circumstances are, of course, something that complicates the discussion of what is at stake when we talk about threats to biodiversity and the necessity of protecting it (Agger and Sandøe 1997, Arler 2009).

Besides the difficulties in defining what a species is, there is a considerable uncertainty related to the question of how many species there have been, how many there are and how fast they are disappearing. At present 1.8 million species are described and preserved according to international taxonomic standards and thereby recognised by the scientific community. However, estimates – more correctly, 'guesstimates' – based on data concerning the rate at which new species are discovered and the thoroughness with which different areas are being researched put the total number of existing species at between 5 million and 100 million species, 13 million species being one of the latest estimates from the United Nations Environmental Programme (UNEP 1997). Thus, according to the most conservative estimate, only a third of the earth's species are known.

The third factor that complicates the debate on the importance of biodiversity is the concept of indigenous species. These are defined as species found within their natural area of distribution, which also is named the biocoenosis they belong to and have adapted to. In this sense, European rabbits in Australia or North American Sitka spruce in Europe are not indigenous species in Australia and Europe respectively. But in many other cases, it is not at all that clear. Just one example that has to do with space and time should be mentioned. Norwegian spruce is the most common tree in Denmark to where it was introduced some hundred years ago. So it is not indigenous, but it is in adjacent southern Sweden. So if we instead of introducing it had waited another hundred or more years it might have appeared by itself and thus been regarded as indigenous.

In densely populated and intensive cultivated landscapes, species are being brought from one place to another either deliberately or accidentally, and human impact changes the environment in many ways that helps the introduced species to survive outside their biocoenosis. In this way there is now, in most cultivated landscapes, a disturbing contrast between the number of alien plant species and the number of indigenous plant species, which are by far outnumbered. Biodiversity may be negatively affected by the disappearance of some indigenous species, but if they are outnumbered by newly introduced species what is then the problem?

The answer is that globally we have many 'local' species and relatively few global species, and it is too often the same relatively few global species that are introduced everywhere at the expense of the very many local species. The problem is comparable with the 'McDonaldization' that hits the towns all over the world. Although the local gastronomic environment now for a period of time may gain in diversity (before the local restaurants are outstripped), we simultaneously lose diversity at a higher level, because it is the same handful of burger chains that expand all around the globe. In the same way, some useful, popular or opportunistic plant species are being spread all around the globe, thus making a short-term addition to local diversity while threatening long-term local and global diversity.

Sometimes an introduced species, after a period of time, may begin to expand aggressively. Such species are called invasive species. And the phenomenon is, in addition to habitat loss and climate change, ranked as one of the major threats to the global biodiversity. The concerns related to the deliberate release of GMOs is a fear that some of them might behave like invasive species.

Sustainable Development and Biodiversity

In the debate on ecology, ideas of a natural balance formerly played a pronounced role, e.g. the Limits to Growth approach argues for the achievement of a state of equilibrium (Meadows et al.1972). Although critical loads and carrying capacity still give meaning, natural balance or natural state of our surroundings have less focus today. An ecosystem may have many points of balance and we, more than 'nature itself', determine which of these points it should be at. In the same way, we formerly paid much attention to biodiversity and its possible influence on the balance of the ecosystem (e.g. Odum 1971). From a biodiversity viewpoint it seems today less relevant. This is not least the case in intensively used agricultural and semi-urban landscapes where most people live. Therefore seen in this narrow context, biodiversity and natural ecosystems become more a result of, rather than a precondition for, sustainable development, whereas in the broader perspective it is the other way around.

In its 'Millennium Ecosystem Assessment', the UNEP classifies the importance of natural ecosystems and the biodiversity they contain in four overarching classes, as given in the list below.

The importance of natural ecosystems may be classified as four different kinds of services (UNEP 2003):

- Provisioning services where products are obtained from the ecosystem like food, fresh water, fuel wood, fibres, biochemicals and genetic resources.
- Regulation services where benefits are obtained from regulation of ecosystem processes like climate regulation, disease regulation, water regulation, water purification and pollination.
- Cultural services where nonmaterial benefits are obtained from ecosystems like spiritual and religious services, recreation and ecotourism, aesthetic, inspirational, educational, sense of place and cultural heritage.
- Supporting services necessary for the production of all other ecosystem services: soil formation, nutrient cycling and primary production (photosynthesis).

Although biodiversity is an integral part of all ecosystems and thus all ecosystem services, it is so in varying degrees. For example, it is possible to substitute most of the provisioning services by cultivating relatively few domestic species and many of the other services can do without a complete and undisturbed biodiversity. Three aspects should, however, be underlined.

A first answer to the question of why the conservation of biodiversity is important for the future, and hence sustainable development, is its role as a source of information. The living organisms (and to some extent also their dead remnants) contain an invaluable accumulated knowledge, i.e. experience of what the living beings and their predecessors have 'found out' while they have adapted to their habitats. And this 'knowledge' may be of significance in relation to the four services mentioned in the bullit list at the start of this chapter. This information is bound to genetic and epigenetic matter in the cells and also to some extent knowledge learned and transferred from one generation to another. If this should be studied and further explored in research and teaching it is obvious that all kinds of species should be kept in one way or another. In theory they could be preserved in one large zoo and botanical garden or better in one or a number of large self-sustained nature reserves around the world, e.g. placed in the 25 biodiversity 'hot spots' that have been localised by the World Wide Fund for Nature (WWF) and the International Union for the Conservation of Nature (IUCN) (Williams 1998). But, in reality, it will only work if it is also distributed extensively outside the 'hot spots'.

The second answer is, however, more demanding; it concerns the recreational value of nature and biodiversity. If biodiversity should play any significant role in people's lives and well-being it has to be distributed all over the globe where people are living. This is not a case of having the same species everywhere, but the indigenous species should be present within their respective natural biocoenosis. This is a claim for giving the biodiversity and the natural settings a widespread and important role as a part of the human environment. Nature and access to natural areas including nearby urban nature is one of the most valued assets in Western societies as is indicated by the prices of real estate adjacent to the seaside or the edges of forests.

A third kind of an answer is that biodiversity, being an integral part of all ecosystems, has an incalculable importance for all or most kinds of services, e.g. as products from the forest or as pollinators or objects of admiration. In particular, many scientists are convinced that a high content of indigenous species gives the ecosystem a high degree of robustness, i.e. the ability to recover after a disturbance. This is because the bio diverse system has a higher degree of adaptability, i.e. a higher preparedness to change whether due to pollution, disturbance of any kind or change climate (Odum 1971). As stated, we only know a tiny fraction of all existing species and even less of all the multitude of relations that exist among them. Therefore, it is a strong argument that it is better to leave the world as it is, instead of letting one species or function disappear after the other. The complexity of the ecosystems and the limitation of our knowledge is an urge to precaution and are, as we will come back to, two of the main motives for being reluctant to release GMOs deliberately.

Thus, it can be concluded that maintaining high or undisturbed biodiversity is not an absolute necessity for every activity and every bit of land of the globe. But there is a strong argument to preserve the total amount of existing biodiversity and the information it contains in one way or another. And there is an insurmountable argument as to why the maintenance of a, as far as possible, undisturbed indigenous biodiversity – not all over but within each land and seascape is important for the protection of the regulation, cultural and supporting services. Cultural services not least have much to do with the ethical aspects of sustainable development that we shall now return to.

The Rise in the Council of the Discussion on Ethical Aspects

In September 2005, the then Danish Minister of Environment, Connie Hedegaard, wrote a letter to The Danish Council of Ethics, asking the Council to provide a statement concerning the concept of 'utility' seen in relation to research in and application of GMOs. Procedures taking 'benefit' as a criterion for approval have existed in varying degrees of success in our neighbouring countries for some years, but the introduction would be new in a Danish context (Kamara and Coff 2006).

It is obvious that the debate relating to GMPs raises much broader issues than those covered by a natural scientific risk analysis. It relates to normative questions that are most often decisive for personal positioning. The Minister expressed her interest in a clarification of 'the more intangible issues that mean so much in the public debate' and have more to do with benefit, ethics and faith than with risk assessment (the distinction between benefit and ethics in two separate categories stems from the Minister's letter). The Minister referred especially to the debate on the application of gene technology in food production. The Council decided to focus on the deliberate release and benefit of GMPs because GMOs are already used to a great extent in other contexts seemingly without any large controversies, e.g. use of genetically modified bacteria in the production of industrial enzymes and medicine (Danish Council of Ethics 2006).[2]

The inspiration might have come from Norway, where the 1993 Gene Technology Act includes criteria of social utility (or 'benefit to the community') and sustainable development as conditions for approval of GMOs for deliberate release (Lov om framstilling og bruk av genmodifiserte organismer 1993). The permission to release depends on utility, and the health and ecological issues which preceding experiments, and consequence and risk analysis have raised. The

---

2   The present chapter is in debt to this work in which one of the authors, Peder Agger, also took part, although the viewpoints expressed in the text presented here are his and not necessarily the Council's. The publication from 2006 is in Danish, but exists also in an English translation: Danish Council of Ethics 2007.

product, the production process and the use are evaluated. According to the Act, the Norwegian Biotechnology Advisory Board ('Bioteknologinemda') must give its comments on all questions related to the Act, including cases of approval of GMOs. The Board assesses whether the release of a GMO will meet the criterion of social utility and further sustainable development. The Board tries to answer questions like: Is there a demand for the product? Will it contribute to relieving a societal problem? Is it better than comparable products? Are there alternatives?[3]

Over the years, the Norwegian Board has frequently concluded that a particular GMO did not meet these criteria but its recommendations have not always been followed by the Norwegian Government. Recently, the Board has responded negatively to three applications for approval of GMPs, whereas the Directorate for Nature Management has been positive (Bioteknologinemda 2008). These cases currently await a decision by the Ministry of Environment, and the conflict demonstrates the contested nature of assessments regarding social utility and sustainability.

Although they are not so far-reaching as the Norwegian example, criteria of utility had already been in use for some years within adjacent areas of legislation. Within the regulation of chemicals REACH (Registration, Evaluation, Authorisation and Restriction of CHemical substances), EU member countries are supposed to use a principle of substitution which requires that the most harmful substances must be avoided if less harmful alternative substances exist. Also, in national legislation on cloning and gene modification of animals it is required that the product should be of clear benefit for health and the environment.

The Concepts of Risk and Benefit

Until now, in Denmark, only considerations of free trade on the world market and the outcome of a natural scientific risk assessment have determined whether a GMP should be released or marketed. It seems, however, that risk is only one among several elements that ought to be considered in the decision (Danish Council of Ethics 2006).

It is, however, not that simple. One reason is that assessments of utility are not always based on the same values for all people. Assessments of benefits include a subjective element. So what is useful? The most obvious answer might be that it is what makes people's lives happier in accordance with Jeremy Bentham's classical principle. But people are very different and have different and often at the same time mutually contradictory wishes. So general standards for what is 'the greatest happiness' or consensus on introducing new technologies, e.g. GMOs, does not seem likely. This, however, does not exclude the possibility of a high degree of consensus on some aspects of the issue.

---

3   See for an account of the background for the Norwegian legislation Hviid Nielsen et al. (2001).

The outcome of benefit assessments also depends on the perspective. What might seem beneficial from an individual perspective often does not from a more social or holistic perspective. The same goes for risk assessments that might be seen as inverted benefit assessments because they deal with negative benefits. Such contradictions point to the need for an open dialogue, e.g. in an advisory board, where different viewpoints might be unveiled and clarified in relation to each other. Such a dialogue is necessary for a democratic legitimation of the final decision which might be formalised so that individual countries could have the right to limit free trade by preventing the introduction of certain GMOs on a culturally determined value basis (Danish Council of Ethics 2006).[4]

Confronted with the difficulties of describing what benefit based on happiness is, many authors have proposed that it should be based on the basic human needs. What they are is, however, far from clear. They cannot be reduced to what might be labelled as essential biological needs, i.e. what a human body needs to survive like food, water and shelter. We are not only living things but also social beings, and as Karl Marx is often quoted as saying, it is not the same starvation that is satisfied with claw and teeth as the one that is satisfied with knife and fork. It might be difficult to define what these needs are. They vary from time to time and from society to society. Still, it is more beneficial to satisfy them than just to satisfy preferences in general. Basic needs have limits depending on the historical and geographical context, whereas preferences in a market society have no limits. Basic needs may be identified through democratic processes and articulated, e.g. in a form like human rights: 'adequate conditions of life, in an environment of a quality that permits a life of dignity and well-being', as it was expressed at the Stockholm Conference 1972 (UN 1972). It is possible in Denmark to establish some kind of consensus on research and development aimed at using GMOs to satisfy basic human needs in this sense.

It might, however, be questioned whether the concept of utility ought to have a dominant role. Some will say that we operate with several ethical principles that have nothing to do with utility, e.g. principles we follow irrespective of whether they are thought to have good or bad consequences. An example could be the principle of preserving every existing plant and animal species – even the most threatening ones like the smallpox virus. On the other hand, it may be argued that such ethical principles are accepted because it is often believed that some of the general consequences are beneficial or at the end have some general beneficial consequences (Danish Council of Ethics 2006). Ethical principles like religious taboos might, for example, often have an ecological bearing.

---

4   With the present legislation this is not possible for the member states but only for sub-governments. 'According to Greenpeace and Friends of the Earth, in EU there are no less than 172 GM-free zones declared by democratically elected administrations and no less than 4,500 local governments that are calling for restrictions on commercialised GM cultivation' (Kamara and Coff 2006). The EU Directive 1829/2003, article 42 allows advice on ethical issues to be obtained from the European Group of Ethics (EGE) or other appropriate bodies of the Commission.

Discussions about utility should not be detached from discussions of risks. Without going deeply into this matter, it seems obvious to us that risks and risk assessments are often taken as simpler and easier than they should be. Kamara and Coff (2006) refer to Brian Wynne's distinction of five levels of uncertainty (Wynne 2006a).

Brian Wynne's different qualities of uncertainty:

1. Risks where we know the probabilities as well as the consequences.
2. Uncertainty where we know the possible consequences but not the probabilities.
3. Ignorance where we don't know the possible consequences nor the probabilities.
4. Indeterminacy where the involved processes are not subject to consistent, predictable outcomes from same initial conditions.
5. Ambiguity where there are differences of meaning among scientists.

Examples of the first type of uncertainty are all the cases in everyday life where cost-benefit considerations can tell us what to do. Examples on the second type of uncertainty may be situations where further research and sampling might reduce the uncertainty to type one. Type three is a situation in which we are doing something without knowing what may happen and that it may happen. Type four is where we really do not know what is up and down. And type five is a well-known phenomenon in the field between science and policy.

Most people will think that only the first type of uncertainty exists, and if the second appears they think we should be cautious and perhaps abstain, which wrongly is understood as using the precautionary principle. This principle, however, should be used if uncertainty three and four appear. Quality five is better known from daily life and is often used to give politicians an excuse to abstain from or postpone any action being taken.

To sum up: if utility should be applied as a criterion for approval of GMPs, a dialogue, possibly formalised in an advisory board and staffed with different kinds of people, could be established. Here discussions can take place on usefulness and what should be considered as relevant human needs and what kinds and level of risks should be accepted. Inspiration can be taken from the home page of the Norwegian Committee of Bioethics, where dozens of cases are presented (in Norwegian: http://www.bion.no).

Reasons for the Minister's Letter

There are several circumstances that may explain the Minister's request. Three of them will be presented here. The first is the recent changes in EU regulation. From 1999 to 2003, a group of five member countries had effectively maintained a moratorium for any new GMPs to be marketed or released. Initially the five countries demanded that new rules regarding the deliberate release of GMOs should be in place before further approvals would be acceptable. But following the passing of the new Directive 18/2001 they further demanded that a proper labelling and control system should be installed first. The EU Directives 1829/2003, 1830/2003 and 65/2004 have implemented this, in Denmark in 2004 (Bekendtgørelse om grænseoverskridende overførsel af genetisk modificerede organismer (Cartagena-protekollen om biosikkerhed og om sporbarhed og mærkning af genetisk modificerede organismer 2004)). The moratorium was then lifted. The regulation implies that the deliberate release of GMPs can be approved after a risk analysis and a public hearing have been carried out (Bekendtgørelse om godkendelse af udsætning i miljøet af genetisk modificerede organismer 2002). The analysis is made in expert committees that examine labelling, traceability and co-existence. In general, the system for approval is of a very technical-scientific character and rather restrictive compared to e.g. chemicals, but it may be justified by the fact that for the dispersal of reproducing unacceptable organisms, the threshold level is zero.

The level of risk acceptance is still open for discussion and disagreement. Are the buffer zones that should keep cultivated GMPs and organic grown fields apart wide enough? And are they at all realistically applicable to the agricultural practice? The pattern of small fields of the Danish landscape makes it very fragmented and thus difficult to control by buffer zones (Quist, Lundsgaard and Brandt 2005) and farmers' practice often deviates from the prescriptions given by the producer or the government. Lassen et al. (2007) have shown that when asked, Danish farmers prefer to use pesticides more frequently than what the producers of pesticide tolerant GMPs prescribe. And a further question arises: Is the threshold (0.9 per cent) of how much a product may contain of GMPs without labelling it acceptable for the consumer? These questions are handled in a current evaluation process (Tolstrup et al. 2007). But this is seemingly not sufficient to remove the public's suspicion. In this situation, an introduction of a criterion of utility could give the Minister a means to avoid the most problematic releases and thus make the administration more flexible and less unpopular.

Another reason for the request could have been the remarkable expansion the cultivation of GMP crops has had globally. The total area has increased continuously from practically nothing in 1996 to 125 million in 2008 (ISAAA Brief No 39 2008). Soybeans, maize and cotton account for almost 90 per cent of the total area covered with GMPs.

A third reason is probably the widespread reluctance in the European population to accept GMOs as food. The latest Eurobarometer from 2005 (Gaskell et al. 2006) shows that although the European population in 2005 was more optimistic and more informed about biotechnology than in any of the five previous times Eurobarometer surveys have been conducted (1991, 1993, 1996, 1999 and 2002), the resistance to GM food is an exception: 'Overall Europeans think that GM food should not be encouraged. GM food is widely seen as not being beneficial, as morally unacceptable and as a risk for society'. The most persuasive reasons relate to health, the reduction of pesticide residues and environmental impacts (Gaskell et al. 2006).

The Eurobarometer states, 'With a few exceptions, among the former EU15 countries we see the tendency of a steady decline in support between 1996 and 1999, and an increase between 1999 and 2002, and a return to a decline in support in 2005'. And in all 15 countries, the level of support in 2005 dropped below those reported in 1996. Contrary to a persistent belief among many scientists and decision makers, the scepticism towards GM foods cannot be seen as an expression of lack of knowledge about modern biotechnology among laypeople (the so-called knowledge deficit model). Eurobarometer results consistently show that there is no connection between knowledge and attitudes to GMOs (see, for instance, Gaskell et al. 2001). Furthermore, when asked about their judgements about various applications of modern biotechnology, respondents all over Europe consistently gave priority to moral acceptability as a predictor for acceptance of an application.[5] A study published in July 2007 showed that now 21 per cent are positive towards and 67 per cent (the highest ever – 58 per cent of the men and 76 per cent of the women) of the adult Danish population are against the use of GMPs in food production (Lassen and Dahl 2007).

It is interesting that although the majority of the citizens are willing to delegate responsibility for new technologies to experts and their decisions based on scientific evidence, a substantial minority would like to see greater weight given to moral and ethical considerations. 'To build further confidence in science policy it would seem prudent to ensure that moral and ethical considerations and the public voice(s) are seen to inform discussions and decisions' (Gaskell et al. 2006).

The regulatory ban on the marketing and release of GMPs has been lifted. The expanding growth of such plants in other countries is knocking on the door. But public reluctance is undiminished – no wonder the Minister felt a need for advice.

The Council's Discussions

During the work on the report, the Danish Council of Ethics has ascertained that sustainability is an overarching consideration that occupies a prominent position in legislation and treaties, also where GMPs are concerned. Before it can be applied in practice, the various considerations covered by the concept have to be weighed up. Some members of the Council therefore think that if a decision concerning a possible release is to be subjected to a sustainability assessment, it should be made by some authority whose make-up is not confined purely to politicians and civil servants. Based on a Norwegian model, such a body might possibly have a more comprehensive mandate to advise and generate debate around the problem issues involved in approving GMPs.

---

5 However, in the Eurobarometer questionnaires on biotechnology the question about moral acceptability was separated from a question about usefulness of the various applications, thus making the same, problematic, distinction between utility and ethics as in the letter to the Danish Council of Ethics from the Minister of the Environment, cf. above.

All members stress the risk that the release of GM species may, owing to the risk of spread, potentially lead to a reduction in biodiversity and ecosystems' ability to function optimally. This risk, combined with the possible irreversibility of such changes, should result in people adopting a restrictive construction of the precautionary principle out of deference for the concept of sustainability and therefore displaying all round reticence as regards the release of GMPs. At the same time, however, the Council takes the view that the controlled release of GMPs should not be excluded a priori that the use of GMPs can contribute to more sustainable and efficient agricultural practice and greater natural abundance (Danish Council of Ethics 2007).

Prior to the drawing of any conclusions, the Council of Ethics discussed three issues that are often discussed in relation to GMPs and their eventual release: Is gene modification significantly different from what we have been doing for centuries? Can the use of GMPs help us to bring down the pollution load? And can GMPs solve the problems of starvation in the Third World?

The introduction of GM is not new in the sense that man – as long as he has had domesticated plants and animals – has performed selection for the purpose of the best offspring. It has been a long, but over the millennia and centuries, very effective technique that has improved the efficiency of domesticated organisms. The new things are that: (A) the GM technique is by its proponents said to be more precise and fast; and (B) genetic material now can be transferred from one species to another – perhaps very different species – a crossing that could never happen outside the laboratory.

As we will see later, opinions in the Council are divided into four lines of thought: Some members think that there are no new reasons for special concern. Others think there are. And some of these members put special weight on the speed of possible change (ecosystems and society with its institutions may not be able to adapt effectively enough), and finally others put weight on the artefact and challenge to what they consider the Lord once created.

All council members agree that GMPs can help to increase production and thus also food production for starving people in Third World countries. But they disagree in how relevant it is to bring this matter up in the present context. The one side points to previous successes and to the present growth in the area with GMP crops. The other side of the debate argues that a phenomenon such as starvation is a much more complicated problema to which there is not any easy technological fix. And the situation, for example in Argentina, demonstrates how the introduction of GMPs by multinational firms has been combined with drastic and negative consequences for the rural population (Joensen 2006).

So the opinions of the Council were also divided on this matter. Some found that the discussion on the possible introduction of the concept of 'utility' in the administration of GMs is meaningless if considerations on the violent changes the introduction of patented GMPs may have on the global agriculture are not included. Others found that this question is important but outside the agenda set by the Minister's letter. And the rest of the Council found that the patent perspective is not anything special for the use of GMOs and therefore not relevant to bring up at all. These discrepancies are intertwined with how members conceive the concept of utility. If it is understood as something that improves human life or happiness, then it is decisive whether it is from an individual or a societal perspective. Some members find that the market will be able to show what is beneficial (profitable). Other members point to the insufficiency of the market because many benefits are not marketable and many 'customers' have no access to the market either because they are too poor or they are not yet born.

The third issue brought up was the possible benefit the use of GMPs might have for the environment. Many of the new GMPs that have been introduced are plants that have been made tolerant to a given herbicide, and the combined producer (of the GMP and the herbicide) claims that the amount of herbicides needed will go down compared to growing of the non-modified plant it replaces. Whether this really happens when it comes to practice has been widely debated among experts, in the public and in the Council. So the present documentation points in both directions, which was enough for some of the council members to show that 'the green argument' for GMPs does not hold true, and the GMP can thus not be seen as useful in this respect. And later on, this view has been supported by an investigation, already referred to above, of attitudes among Danish farmers (Lassen et al. 2007).

The Council's Conclusions

The Danish Council of Ethics does not see it as its task to reach a consensus, but to unfold the dilemmas, and elucidate the premises and possible arguments that ought to be elements of a decision taken by politicians or others. After having unveiled a number of basic ways of formulating the problem(s), in order to create an overview, the Council presented three main positions to the question of whether utility ought to be a criterion for the approval of GMPs.

Initially, consensus was expressed concerning the need to start by assessing the possible risks and that research in GMPs can be seen independently of eventual benefit by deliberate release of GMPs. Independent and critical research at international level is seen as decisive for taking a responsible position on an eventual use of GMOs (Danish Council of Ethics 2006).

Further, it was agreed that there is a risk that dispersal of modified genes may lead to a decrease in biodiversity and the ability of ecosystems' optimal function. This may either be due to the spread of alien genes/plants which may compete with indigenous genes/plants in the wild, or it may be caused by the uniformities of cultivated strains of crops. Combined with the threat that the changes can be irreversible, we ought to apply a restrictive interpretation of the precautionary principle, with reference to the concept of sustainability, and generally be restrictive in the release of GMPs. Although it is stated that the Council does not find that a controlled deliberate release of GMPs can be excluded beforehand.

It was maintained that an assessment of whether a given GMP should be released or not ideally has to be made holistically. The assessment should, e.g. consider the possibility of irreversible consequences, and the concept of sustainable development should be a part of the normative basis for a decision in the sense that it should include considerations for future generations and those in poverty. But the Council also pointed to the multifaceted character of the concept and warned that using the concept of sustainability in the debate about GMPs could make the debate diffuse, because a formal agreement about the importance of a sustainable development could cover up conflicting views on different priorities in relation to a sustainable development. Despite the many possible interpretations of the concept of sustainability, the Council nevertheless agreed that the concept has and has had important and very positive implications for the debate about and the regulation of GMPs, in particular because it has contributed to giving ethical considerations a prominent position in the debate (Danish Council of Ethics 2006).

Based on mutual agreement on these points, a majority found that benefit ought to be considered in a governmental procedure for approval. Some of these members think that GMPs should only be released if it can be assured with high probability that there is only a minimal risk for negative effects and there is a fair chance that it will be proven useful, i.e. either is able to satisfy basic human needs or abate serious environmental problems. For some, their position has a religious foundation. Others refer to an inherent 'wisdom' accumulated in nature, which it is not wise to challenge more than strictly needed. But others found that GMPs may be approved if the assessment turns out to show that the expected benefits clearly outbalance the drawbacks. And they also find that GMPs may be accepted even if the usefulness is insignificant as long as the risks are deemed to be insignificant. The minority found that considerations about the benefit or usefulness should be a part of the approval procedure. If it can be argued that humans or the environment will be at risk, they find that any release of a GMP should be excluded. And if it is not the case, then the plant should be allowed without further complications.

Reflections on Acceptance and Resistance

We cannot in any sensible way recount the whole debate in the Council that led to this outcome. But two arguments should be mentioned. One is related to the discussion about risk assessment. All members found, as mentioned already, that a risk assessment should be carried out before any approval can be given. It is, however, our impression that some members have demonstrated a misconception of what risks are and what risk assessments may bring. Firstly, they have what we will call an undue and over confidence in how precise and sufficient risk assessments can be made. It is our impression that they are not aware of the kinds of uncertainty that Wynne termed ignorance, indeterminacy and ambiguity. Secondly, they might not be aware of the normative character of any assessment. Risk assessments are never strictly objective, but always 'tainted' by norms, e.g. in the confinement of what is at all considered. Risk assessments are what Ulrich Beck called 'implicit ethics' (Beck 1992, Jelsøe 2006a). Gaskell et al. (2007) underline this by comparing what they call two cultures of risk: the scientific definition of risk as 'the probability of an unacceptable loss' and the 'intuitive understanding' where risk 'is a complex, socially narrated concept based on a variety of factors, cognitive and non-cognitive'.

Another argument that needs mentioning concerns who will be the best to judge what is beneficial and what is not. The majority found that some kind of collective institution, like the Council itself, could provide the answers. But a significant minority found that the best judge of whether a new product is useful or not is the consumer. These council members found, therefore, that the market, and not a council, should be the mechanism that limits or opens to a new product.

As pointed out by Coff (2006b): eating is for us a very intimate contact to the rest of the world. It implies incorporation, digestion and (re)incarnation, and is often important for the cultural and individual construction of identity. 'You are what you eat' underlines, metaphorically, the relationship. In contrast to drugs, food is both culture and identity. This is probably a part of the explanation why GMO food is met with so much more reluctance than other GMOs. Especially provocative in the present regulation is the incomplete labelling that hides GMO concentrations below one per cent and a regulation that permits meat from GMP fed husbandry to be marketed without labelling. The health of both your body and soul and the health of the outer environment play a role in varying degrees when an increasing number of people buy organic food.

Still another source of scepticism is the series of scandals we have seen within European agriculture over the last decade: mad cow disease, foot-and-mouth disease, the dioxin scandal, the illegal and legal use of growth hormones, straw shorteners, animal welfare problems, and all the accusations for polluting the environment with nutrients and pesticides, hormones, etc. These problems have had a detrimental effect on the reputation of conventional farming and the food sector, which has been weakened further by business behaviours such as tough competition within the sector that is found ethically dubious by the public at large. This leaves an impression among many people of an irresponsible and untrustworthy industry that may easily lie about possible risks and take chances on behalf of the consumer in the continuous hunt for profit.

The third explanation also has to do with reputation, but this time among companies producing the GMPs. Monsanto has gained a really bad reputation due to the way in which the firm has introduced its products, e.g. Round-Up-Ready-tolerant crops in Europe (Nottingham 2002) as all over the world (Joensen and Semino 2004, Joensen 2006). Again this is because of crude business methods, but it is also caused by unconvincing results seen from the consumers' position. What might he or she benefit from sugar beets that have been sprayed with Round-Up? The answer is 'nothing' or almost nothing because the profit goes first to Monsanto and next to the farmer.

The proponents of GM technologies have mistakenly thought that consumers think like rational scientists or economists. But consumers are not only consumers but also citizens. And besides being what Gaskell et al. call 'intuitive scientists or economists' they are also 'intuitive politicians' and 'intuitive ethicists' concerned with fairness, social interests and justice as well as existential questions and values (Gaskell et al. 2007).

What we find is an expression of a broader cultural resistance that is fuelled by many other things than the release of GMOs as such. It may, for example, be resistance against multinational companies and imperialism or against the modern high-tech discourse for development. This phenomenon is well known from other conflicts between environmental movements and the establishment where a broader discontent is triggered by something more specific, e.g. the resistance against the reintroduction of wolf in the Norwegian mountain areas (Skoogen 2003), and the resistance against nature restoration and conservation in Denmark (Jensen and Hansen 2007).

In general, and that is the last reason that is presented here, the debate suggests that we are at the beginning. From the industry there are many promises but few proofs. The industry promises enrichment but it has often ended in impoverishment. Almost all of the first generation of GMPs that have been released until now have been plants tolerant to specific herbicide products or plants made resistant to pests by inoperation of genes from Bacillus thurigiensis (Steinbrecher 2001). None of these have been of self-evident benefit to the consumer. But at the same time we have still not seen any of the worst-case scenarios that the sceptical participants in the debate have envisioned.

Where the first generation of GMPs thus have mainly dealt with the production process, the next generation may deal with the quality of the product like nutritional value, flavour and functional food, and the third generation promises completely new functions of plants as growing of pharmaceuticals (Coff 2006b). Whether this will change the reputation of the GMOs remains to be seen.

The experience with the introduction of and debate on the effects of GMOs has several analogies with the much 'larger' debate on climate change. Both concern environmental issues and thus highly exposed to doubt and uncertainties due to the complexity of the issue and the irreversibility of its effects. This contributes to a deepening of the cleft between experience-based layman knowledge and scientific-based expert knowledge. To this can be added a considerable mistrust because commercial and/or political interests are active behind the screen, at times to a degree where these interests take over so biodiversity or the climate are taken as hostages in a game which has other more short-sighted economical interests.

For both themes, considerations about sustainable development are relevant and helpful in a search for solutions: Needs for the poor and for future generations ought to be considered along with fairness in the distribution of burdens, and maintenance of natural resources and nature as necessities if a revitalisation of economy should be made possible. In both cases things have to be more concrete before any sustainable development might be implemented.

Scientists and Citizens

It is obvious that the conditions for a political debate on the deliberate release of GMOs are changing. The stop of the moratorium on approval of releases in the EU, the increasing area of GMP crops outside the EU and the first steps to grow it inside the EU encourage the industry and its followers. Public opinions have been changing in three ways. One is an increasing acceptance of all kinds of GMOs but food-related applications. The other is the persistent, and in some countries even increasing, resistance to the release of genetically modified food crops. And the third is the greater weight that is being given to moral and ethical considerations. This is demonstrated both in the Eurobarometer surveys and by a growing number of policy initiatives and deliberations regarding ethical implications of food and biotechnology not only in Denmark but internationally (as a recent example in the European context, see EGE 2008).

We agree with the more sceptical side of the Council requiring the risk to be at minimum and the benefits convincing before any release should be accepted and that using sustainable development as a criterion for the utility of GMPs will be appropriate. However, in this connection it is important to emphasise the value foundation of assessments of both risk and ethics. Furthermore, risk and ethics in many situations are intimately connected so taking utility arguments into consideration for the approval of GMOs will likely highlight the contested and politically and culturally embedded nature of both risk and ethics. This points to the need for a broad public debate of these matters and raises the question of what we and all other equally sceptical experts as well as laypeople can do.

For the researcher, it is important to be both explicit and at the same time reflexive, both about the subjectivism described in the example above and in the sense that it should be possible for others to look over one's shoulder in order to learn about one's premises. This requires that these, as far as possible, are available, accessible and possible to understand for others than experts. This is a challenge to many scientists given the expert culture in which science as an institution is embedded (Agger 2004, see also Wynne 2001).

Besides supporting what is mentioned already we as laymen and ordinary citizens can claim influence through advisory boards, public hearings, public access to information and require higher levels of safety. Yet, as a number of studies have shown, it is important that such institutional arrangements to a higher degree enable a critical focus on the different cultures of risk and a broader perspective on new technologies than that of risk assessment. In that connection, the question about the utility of new biotechnologies is relevant. An additional question in connection with this, which we cannot go further into in this connection, however, is at what stage in the process citizens are involved, i.e. in the early stages of innovation or when new and already developed products are in the phase of approval (Wilsdon and Willis 2004, Wynne 2006b). As consumers we can also exert our influence via the market, e.g. buy organic food and claim that the further distribution of organically produced products should be facilitated or not – buy GM food (except for the fact that in many European countries it is still virtually impossible to buy food products which are labelled as GMO as a reflection of the consumers' resistance towards GM foods).[6]

Conclusion

The comprehension of gene technology, like climate change, is fenced in by uncertainty and complexity, and it is a technology to which there are both high positive and negative expectations and thus also disagreements about. Natural scientific risk assessment is obviously challenged by other risk conceptions, e.g. among laypeople and experts from other disciplines. Therefore it can be more constructive to look at the different conceptions as social constructions. Things like trust (or lack of it) and people's thoughts as intuitive politicians and ethicists should, therefore, be considered seriously along with the more traditional natural scientific rationality.

The proposals about including utility and, in connection with that, sustainable development as criteria for approval of GMPs are positive in this regard, since it necessarily implies a broader and more proactive view on the need for new technologies involving considerations about social needs as well as the utilisation of natural resources. More specifically, the questions posed in this chapter are on four levels:

- The first question is whether GMPs should be released at all? Our answer to that question is conditional in the sense that there is an obvious need for institutional arrangements that can accommodate the normative character of assessments of both risk and utility.
- The second question is concerned with the possible inclusion of utility as a criterion for approval of GMOs. We find this to be a good idea because it will give a broader basis for a societal debate about GMOs. Furthermore, when dealing with genetic technologies it seems well-founded to assess the utility as a basis for approval, just like it is done in relation to human biotechnologies.
- The third question, is sustainable development, as promised at the beginning, a normative concept that really helps us to organise politics at a higher level? We think it has been helpful in the present context. The integrative character of the concept, including both human needs and

---

6 The fact that farm animals in Europe to a considerable extent are fed with feed of GMPs without any labelling, which is perfectly in accordance with the current EU legislation, is an indication of the still existing difficulties in achieving transparency regarding these matters. But organically produced food is devoid of any GMPs even from animal feed according to the rules for organic farming.

natural resources, as well as an inter-generational perspective makes it an appropriate foundation for a societal debate, clarification and decision making.

- Fourthly, does this case story contribute to biodiversity and a new sustainability agenda? Our answer is that the increasing complexity in the global society challenges its democratic institutions in many ways. The case given here demonstrates how moral issues and layman influence become necessities in sustainable development.

As scientists we, of course, should continue our scientific work as a means to describe and understand the true facts of the case. Here, especially, it is important to help others to understand the five different levels of uncertainty that we referred to when presenting Wynne's views. But knowing that even facts are always tainted with subjectivism, if not in other ways then because they are always an element in a societal context (Turner and Wynne 1992), we should also be painfully aware that we are not the only ones who might claim that they have found the truth.

By making clear the normative foundation of the discussions about GMOs on very broadly accepted values, it will be easier to come around the debate in a way, where laypeople's judgements are not too dependent on experts and the experts are freed from the cage of a claimed objectivity. In this way a democratic dialogue is better facilitated and, who knows, there may be a wiser handling of GMOs in the future.

Acknowledgements

Thanks to the colleagues in The Council of Ethics for good discussions, and thanks to Nanna Skriver and Henrik Jørgensen from the secretariat who compiled the report (Danish Council of Ethics 2006).

References

Agger, P. 2002. Between Science and Policy: The Experience of the Danish Nature Council, *The Journal of Transdisciplinary Environmental Studies (TES)*, 1(1), 1–13.

Agger, P. and Sandøe, P. 1997. The Use of 'Red Lists' as an Indicator of Biodiversity, in *Cross-Cultural Protection of Nature and the Environment*, edited by F. Arler and I. Svennevig. Odense: Odense University Press.

Arler, F. 2009. *Biodiversitet – Videnskab, Kultur, Etik, Vol. I–II*. Aalborg: Aalborg Universitetsforlag.

Beck, U. 1992. *Risk Society: Towards a New Modernity*. London; Thousand Oaks; New Delhi: Sage Publications.

Bekendtgørelse om Godkendelse af Udsætning i Miljøet af Genetisk Modificerede Organismer 2002. Nr. 831 af 03/10/2002, Copenhagen.

Bekendtgørelse om Grænseoverskridende Overførsel af Genetisk Modificerede Organismer (Cartagena-protekollen om Biosikkerhed) og om Sporbarhed og Mærkning af Genetisk Modificerede Organismer 2004. Nr. 1153 af 10/11/2004, Copenhagen.

Bioteknologinemda. 1999. *Bærekraft, Samfundsnytte og Etikk i Vurderinger av Genmodifiserede Organismer: Operationalisering av Begrepene i GenteknologiLov om Framstilling og Bruk av Genmodifiserte Organismer ens §§1 og 10*. [Online: Bioteknologinemda]. Available at: http://www.bion.no [accessed: 9 November 2009].

Bioteknolognemda. 2008. *Norske GMO-vurderinger – Kommentarer til Direktoratet for Naturforvaltnings Slutrapporter til Miljøverndepartementet av 2. juni 2008 for Maislinjene T25 og NK603*. [Online: Bioteknologinemda]. Available at: http://www.bion.no/uttalelser.shtml [accessed: 29 April 2009].

Coff, C. 2006a. *The Taste for Ethics: An Ethic of Food Consumption*. Dordrecht: Springer.

Coff, C. 2006b. *GMOs and Sustainability: Contested Visions, Routes and Drivers*, BioCampus Conference: Public Perception of Gene Modified Organisms. Copenhagen: University of Copenhagen, 24 November 2006.

Convention on Biological Diversity. 1992. [Online]. Available at: http://www.cbd.int [accessed: 9 November 2009].

Danish Council of Ethics. 2006. *Nytte, Etik og Tro i Forbindelse med Udsætning af Genmodificerede Planter*. Copenhagen: Danish Council of Ethics. [Online: Danish Council of Ethics]. Available at: http://www.etiskraad.dk [accessed: 9 November 2009].

Danish Council of Ethics. 2007. *Utility, Ethics and Belief in Connection with the Release of Genetically Modified Plants*. [Online: Danish Council of Ethics]. Available at: http://www.etiskraad.dk/sw14174.asp [accessed: 9 November 2009].

EGE. 2008. *Ethics of Modern Developments in Agriculture Technologies. Proceedings of the round-table debate*, Brussels, 18 June 2008. Brussels: Secretariat of the EGE, European Commission.

Gaskell, G., Allansdottir, A., Allum, N., Corchero, C., Fischler, C., Hampel, J., Jackson, J., Kronberger, N., Mejlgaard, N., Revuelta, G., Schreiner, C., Stares, S., Torgersen, H. and Wagner, W. 2006. *Europeans and Biotechnology in 2005: Patterns and Trends. Eurobarometer 64.3. A Report to the European Commission*. [Online]. Available at: http://ec.europa.eu/research/press/2006/pdf/pr1906_eb_64_3_final_report-may2006_en.pdf [accessed: 9 November 2009].

Gaskell, G., Allum, N., Wagner, W., Hviid Nielsen, T., Jelsøe, E., Kohring, M. and Bauer, M. 2001. In the Public Eye: Representations of Biotechnology in Europe, in *Biotechnology 1996–2000: The Years of Controversy*, edited by G. Gaskell and M.W. Bauer. London: Science Museum, 53–79.

Gaskell, G., Kronberger, N., Fischler, C., Hampel, J. and Lassen, J. 2007. *Consumer Perceptions of Food Products from Cloned Animals: A Social Scientific Perspective. Prepared for the European Food Safety Authority*. [Online:

European Food Safety Authority]. Available at: http://www.efsa.europa.eu/en/events/documents/stakeholder080207-p5.pdf [accessed: 9 November 2009].

Hviid Nielsen, T., Haug, T., Berg, S.F. and Monsen, A. 2001. Norway: Biotechnology and Sustainability, in *Biotechnology 1996–2000, the Years of Controversy*, edited by G. Gaskell and M.W. Bauer. London: Science Museum, 237–50.

ISAAA Brief No 39. 2008. *Global Status of Commercialized Biotech/GM Crops: 2008.* [Online]. Available at: http://www.isaaa.org/RESOURCES/publications/briefs/39/default.html [accessed: 9 November 2009].

Jelsøe, E. 2006a. *GMO og Nytte.* Paper for a seminar:Hvordan kan nytte, etik og tro indgå sammen med risikovurderinger i vurderingen af GMO? Arranged by a working group under the Danish Council of Ethics, Copenhagen, 13 January 2006.

Jelsøe, E. 2006b. *Public Perceptions of Biotechnology: The Eurobarometer 2005, BioCampus Conference: Public Perception of Gene Modified Organisms.* Copenhagen: University of Copenhagen, 24 November 2006.

Jensen, C. and Hansen, H.P. 2007. *Arven fra Vadehavet, in Økologisk Modernisering på Dansk*, edited by J. Holm, L.K. Petersen, J. Læssøe, A. Remmen and C.J. Hansen. Copenhagen: Frydenlund, 187–222.

Joensen, L. 2006. *Biotechnology to Help the Poor.* Paper presented at the conference: Om nytte, etik og tro i forbindelse med udsætning af genmodificerede planter, Danish Council of Ethics, Eigtveds Pakhus, Copenhagen, 28 September 2006.

Joensen, L. and Semino, S. 2004. *Argentina: Estudio de Caso sobre el Impacto de la Soja RR.* [Online: Grupo Reflexion Rural]. Available at: http://www.grr.org.ar [accessed: 9 November 2009].

Kamara, M. and Coff, C. 2006. *GMOs and Sustainability: Contested Visions, Routes and Drivers. Report Prepared for the Danish Council of Ethics.* [Online: Danish Council of Ethics]. Available at: http://www.etiskraad.dk/sw11043.asp [accessed: 9 November 2009].

Kemp, P. 2000. *Bæredygtighedens etik, in Dansk Naturpolitik – i Bæredygtighedens Perspektiv*, edited by J. Holten-Andersen, T.N. Pedersen, H. Stensen Christensen and S. Manninen. Copenhagen: Naturrådet, 14–23.

Lassen, J. and Dahl, B. 2007. Hvis bare folk vidste mere ..., *Food Culture*, 2007(25), 6–7.

Lassen, J., Nielsen, D.E., Vestergaard, L. and Sandøe, P. 2007. *Miljøvenlige Genmodificerede Afgrøder? Vil Landmændene have dem, og vil de blive brugt til gavn for Naturen?* [Online: Bekæmpelsesmiddelforskning fra Miljøstyrelsen, Nr. 112]. Available at: http://www2.mst.dk/Udgiv/publikationer/2007/978-87-7052-543-5/pdf/978-87-7052-544-2.pdf [accessed: 9 November 2009].

Lindsay, N., Kamara, M.W., Jelsøe, E. and Mortensen, A.T. 2001. Changing Frames: The Emergence of Ethics in European Policy on Biotechnology, *Notizie di POLITEIA*, XVII(63), 80–93.

Lov om Framstilling og Bruk av Genmodifiserte Organismer m.m. (Genteknologiloven) 1993. Nr. 38, Oslo. [Online]. Available at: http://www. Lov om Framstilling og Bruk av Genmodifiserte Organismer data.no/all/hl-19930402-038.html [accessed: 9 November 2009].

Mayer, E. 1942. *Systematics and the Origin of Species.* New York: Columbia University Press.

Meadows, D.H., Meadows, D.L., Randers, J. and Behrens III, W.W. 1972. *The Limits to Growth.* New York: Universe Books.

Mohr, H. 1987. *Natur und Moral. Ethik in der Biologie.* Darmstadt: Wissenschaftlige Buchgesellschaft.

Norwegian Biotechnology Advisory Board 2006. *Sustainability, Benefit to the Community and Ethics in the Assessment of Genetically Modified Organisms: Implementation of the Concepts set out in Sections 1 and 10 of the Norwegian Gene Technology Act* [Online: The Norwegian Biotechnology Advisory Board]. Available at: http://www.bion.no [accessed: 9 November 2009].

Nottingham, S. 2002. *Genescapes – The Ecology of Genetic Engineering.* London: Zed Books.

Odum, E.P. 1971. *Fundamentals of Ecology.* London: Saunders.

Prip, C., Wind, P. and Jørgensen, H. 1996. *Biological Diversity in Denmark: Status and Strategy.* Copenhagen: Danish Forest and Nature Agency.

Quist, H.W., Lundsgaard, H. and Brandt, J. 2006. GMO Neighbourhoods: Will Co-existence be a Geographically Realistic Possibility? *Geografiska Annaler Series B, Human Geography,* 88B(2), 199–213.

Schutz, J. 1989. Das Gehirn als System. *Naturwissenschaftlige Rundschau,* 42, 345–53.

Skoogen, K. and Krange, O. 2003. A Wolf at the Gate: The Anti-carnivore Alliance and the Symbolic Construction of Community, *Sociologica Ruralis,* 43(3), 309–25.

Steinbrecher, R. 2001. Ecological Consequences of Genetic Engineering, in *Redesigning Life? The Worldwide Challenge to Genetic Engineering,* edited by B. Tokar. London: Zed Books, 75–102.

Tolstrup, K., Andersen, S.B., Boelt, B., Gylling, M., Holm, P.B., Kjellsson, G., Pedersen, S., Østergård, H. and Mikkelsen, S.A. 2007. *Supplerende Rapport fra Udredningsgruppen Vedrørende Sameksistens mellem Genetisk Modificerede, Konventionelle og Økologiske Afgrøder. Opdatering af Udredningen fra 2003* [Online: Ministry of Food, Agriculture and Fisheries, The Danish Plant Directorate]. Available at: http://pdir.fvm.dk/Default.aspx?ID=8960 [accessed: 9 November 2009].

Turner, G. and Wynne, B. 1992. Risk Communication, in *Biotechnology in Public – A Review of Recent Research,* edited by J. Durant. London: Science Museum for the European Federation of Biotechnology, 109–41.

UN. 1972. *Declaration of the United Nations Conference on the Human Environment.* Stockholm [Online: United Nations Environment Programme]. Available at: http://www.unep.org [accessed: 9 November 2009].

UNEP. 1997. *Global Environment Outlook-1, United Nations Environment Programme, Global State of the Environment Report 1997*. [Online: United Nations Environment Programme]. Available at: http://www.unep.org/geo/geo1/ch/toc.htm [accessed: 9 November 2009].

UNEP. 2003. *Ecosystem and Human Well-being: Millennium Ecosystem Assessment*. London: Island Press.

Williams, P.H. 1998. Key Sites for Conservation: Area-selection Methods for Biodiversity, in *Conservation in a Changing World*, edited by G.M. Mace, A. Balmford and J.M Ginsberg. Cambridge: Cambridge University Press, 211–50.

Wilsdon, J. and Willis, R. 2004. *See-through Science: Why Public Engagement Needs to Move Upstream*. London: Demos.

World Commission on Environment and Development. 1987. *Our Common Future*. Oxford: Oxford University Press.

Wynne, B. 2001. Creating Public Alienation: Expert Culture of Risk and Ethics on GMOs, *Science as Culture*, 10(4), 445–81.

Wynne, B. 2006a. *GMO Risk Assessment under Conditions of Biological (and Social) Complexity*, Austrian Government, EU Presidency Conference, Hofburg, Wien, 18–19 April 2006. [Online]. Available at: http://www.umweltbundesamt.at/fileadmin/site/umweltthemen/gentechnik/GMO-Precaution_pdfs/GMOPrec_wynne1_170406.pdf [accessed: 29 April 2009].

Wynne, B. 2006b. Public Engagement as a Means of Restoring Public Trust in Science: Hitting the Notes, but Missing the Music? *Community Genetics*, 9(3), 211–20.

PART II

Sustainability in Relation to Political and Institutional Actions and Activities

Chapter 5

Sustainable Transition of Socio-technical Systems in a Governance Perspective

Ole Erik Hansen, Bent Søndergård and Jens Stærdahl

Introduction

Present patterns of production and consumption are not sustainable. Therefore deliberate action for sustainability demands a focus on the transition of societies' socio-technical systems. This implies a new perspective for sustainable planning and policies as it becomes a question of addressing the governance structures of socio-technical systems, understanding how such governance structures develop and shape technological development. It demands metagovernance, which is the ability and strategies to influence and shape such governance structures. This involves the question of how to stimulate reflexive processes in actor networks in existing or developing socio-technical systems in order to integrate the concept of sustainability as a driver for transition.

Building on this understanding, this chapter sets out to examine how it is possible through deliberate processes to lead the development of production and consumption systems in a more sustainable direction. The main questions that need answering here are: *How should we understand governance structures of socio-technical systems and how can transition strategies concerning such structures be developed?*

Sustainable Transition and Modernisation

The concept of sustainability has been discussed in environmental policies and development strategies for the last 20 years. However, the substantial implementation of sustainable programmes has been limited. Western European countries have witnessed capacity building and the transformation of production within programmes of ecological modernisation. Within this policy scheme, programmes for cleaner technologies and products in combination with the adoption of elements of green strategies in companies have resulted in the development of cleaner productions and products, which have reduced the negative environmental impacts, as well as in the building of distributed environmental capacities in terms of reconfigured actors, institutions and networks (Mol 2000).

As a result, we have experienced a modernisation process based on institutional reflexivity and the development of specific capacities to act in industrial sectors (Søndergård et al. 2004). However, assessed from the perspective of sustainability, the efficiency gains of ecological modernisation programmes are inadequate: gains are often outweighed by subsequent growth in volumes, and environmental problems without a marketable technological solution are not dealt with (Jänicke 2007). More radical transition processes have proved difficult to establish within the framework of an ecological modernisation approach because this approach relies heavily on actors embedded in the rationality of the existing socio-technical systems. Faced with the challenge of sustainability, ecological modernisation programmes have proved insufficient, leaving a need for developing new strategic programmes of sustainable transition capable of supporting more radical transition processes.

Establishing sustainable transition is the challenge of our time and can be compared to the development of democracy and the welfare state; both were (in parallel to sustainability) open concepts subject to interpretations and struggles in earlier stages of modernity (Voß and Kemp 2006). However, in contrast to struggles for democracy and welfare, no specific social agents who 'carry' the process can be identified. Other agendas, such as the neo-liberal agenda and the globalisation agenda, seek to dominate the modernisation process. The transformation towards sustainability in most cases has been subordinated to other agendas or has been neglected. An example is the actual subordination of EU environmental policy to the Lisbon process (making the achievement of European global competitiveness the main argument) or the parallel linking of attempts to orchestrate strategic EU climate responses to policies of energy supply security, business opportunities and crisis management. There is a problem of agency and power, and a contextual problem where modernisation processes (globalisation, individualisation and more) form the context of sustainable transition, establishing 'competing processes of structural transformation' (Grin 2009: 308). Western societies are at a level of material production and consumption at which sustainability and modernisation cannot be separated. Efforts within the lines of ecological modernisation, however, have been shown to have important limitations in relation to providing needed radical transitions. This points to the need for policies of sustainable transition and deliberate action for sustainability.

We aim to contribute to this end, and thus we examine how the idea of 'sustainability' can guide deliberate transitions of socio-technical systems. The first part introduces socio-technical systems, how we may understand their dynamics and the challenges involved in governing their development in a more sustainable direction. The second part addresses transition strategies including Dutch experiences with 'transition management' and the EU's strategies for 'low carbon energy systems'. Finally, the governance of sustainable transition is discussed in a planning perspective.

Sustainable Transition of Socio-technical Systems

How can complex social systems be subject to deliberate transformation? In order to obtain a more profound understanding of the challenges of deliberate transition, we are going to address socio-technical systems as specific path-dependent configurations, socio-technical systems as regimes and governance structures, the dynamics of socio-technical systems conceived as the interplay with socio-political landscapes and path breaking local activities (niches), and the transition of socio-technical systems in a governance perspective.

Socio-technical Systems and Path Dependency

Production and consumption in modern societies can be perceived as being organised in a number of socio-technical systems that fulfil social functions and needs. From a sustainability point of view, some of the crucial socio-technical systems of society are the production and consumption systems of food, of construction and housing, of heat and power, and of transportation (Vellinga 2001). If we are going to establish more sustainable modes of fulfilling social functions and needs, we have to be able to make deliberate changes of socio-technical systems providing those social functions. This makes an understanding of processes of formation, stabilisation, reproduction and the dynamics of changes of socio-technical systems a core issue of environmental and sustainable policy and planning.

Research on the transition of social-technical systems has strong linkages to the social construction of technology approach (Pinch and Bijker 1990) and actor network theory (Callon 1991), building on their understanding of technology as being socially embedded in social networks and shared cognition, and their perception of technology development as processes of reconfiguration of technology, actors, networks and institutions (Geels 2004a, 2004b, Hughes 1993). Technologies are not just technical elements but, functioning in society, they involve linkages between heterogeneous elements combining such components as physical artefacts, manufacturing firms, scientific knowledge, institutions, users, etc.

Following this line of understanding, the individual socio-technical system is understood as specific configurations of such elements as technology, regulation, infrastructure, supply networks, maintenance networks, user practices, markets and cultural meaning (Geels 2004a, 2004b). Car-based mobility can be an illustrative case. Such elements as fuel and road infrastructures, maintenance and distribution networks, mobility patterns and driver preferences, and symbolic meaning (e.g. freedom and individuality) all contribute to constitute and reproduce the socio-technical system (Geels 2005).

What makes the elements into a *system* is the integration into a seamless web (Hughes 1986) or, put differently, all the linkages between the elements that make it possible to fulfil a societal function as transportation (Geels 2004a: 900). In a transition perspective, it implies that all those elements (markets, user practices, regulation, infrastructure, cultural meaning and so on) have to be addressed and involved in a transition process.

Looking at the dynamic of socio-technical systems, technological development is taking place within systems where existing technologies and cognitive and institutional structures frame technological development. Once the systems are established and have gained momentum, innovation normally becomes incremental and dominated by system-optimising innovations within the system (Hughes 1993, Dosi 1982), that is, technology development becomes path dependent (e.g. David 1985, Arthur 1990, Walker 2000) and attempts towards more radical shifts are resisted. Key mechanisms behind this path dependency are systemic linkages, institutions and networks of actors (Geels 2004a).

Systemic linkages relate to functional and economic interdependencies. The systemic nature of production implies that the many sub-components in the system depend on each other, which creates inertia as changing one component requires changes in a host of other components. In economic terms, the sunk costs in established material structures in the systems (e.g. coal-fired power plants impede changes to different development paths, economies of scale) imply that once the production of a specific component has gained the lead over alternative components, the declining costs of mass production will give a competitive advantage (Geels 2004a: 911).

Legal, normative and cognitive *institutions* are important constitutive and path shaping elements of social-technical systems. The regulative institutions such as, for example, standards, laws, subventions, etc. are adjusted to and often privilege the prevailing technologies and hamper the development of new technologies (Walker 2000). In parallel, normative institutions such as mutual expectations between users and producers, and normative commitment amongst policy makers to existing socio-technical systems further privilege existing technologies (Geels 2004a: 910, Walker 2000), and cognitive institutions in technological development make innovators expect and search solutions within the path of existing technologies (Dosi 1982).

Finally, the *networks* of the socio-technical system, the specific configuration of actor relations, which form the framework of innovation and production, induce rigidity. Networks once formed, resist change, as change is costly (Geels 2004a: 910–11) and involve the redefinition of roles and power relations of vested interest and incumbent players.

High complexity is related to these systemic interconnections; the development of socio-technical systems takes place as interaction between many actors, networks and institutions. The fact that socio-technical systems develop in interaction with other complex systems, which co-evolve, add further to this complexity. Complexity makes prediction of the outcome of interventions uncertain. Socio-technical systems defy rational planning; undertaking deliberate transition is better conceived as processes of governance.

Regimes and Governance Structures

In an innovation and transition perspective, the concept of 'regime' has been introduced to emphasise that socio-technical systems are relational unities where technologies, actors, networks and institutions constitute and reproduce specific rationalities guiding innovation practices. *A socio-technical regime* can be defined as a 'semi-coherent set of rules carried by different social groups. By providing orientation and coordination to the activities of relevant actor groups, socio-technical regimes account for the stability of socio-technical systems' (Geels 2004b: 33). Thus, the regime is the 'rules of the game' of a socio-technical system.

The regime perspective emphasises that socio-technical systems are governed and reproduced by the activities of social actors. Socio-technical regimes are complex systems subject to attempts at steering by actors inside and outside the regimes; they are the collective – and contingent – outcome of the strategic choices and social interaction of many actors. Due to this complexity of socio-technical systems, the steering processes of socio-technical systems are best *understood as activities of governance in self-organising inter-organisational networks* (Rhodes 1997: 53). These networks resist government steering and develop their own policies and mould their environment (Ibid. 52).

In relation to the transition of socio-technical systems, we will make a distinction between 'governance within' socio-technical systems and 'governance of' socio-technical systems. Looking at governance within socio-technical systems, governance activities (governance processes, agency) producing and reproducing the system take place in structural settings, governance structures, established as collective outcomes of former distributed governance processes. There is 'a duality of structure' (Giddens 1976); governance structures are the media shaping agency and governance activities and they are the outcome of agency and governance activities. 'Governance structure' is at the centre of understanding the dynamics of socio-technical systems, and is understood as a combination of the aligned institutions and patterns of agency/steering, the material structures in terms of functional and economic interdependencies, and configurations of actors and networks.

Looking at transition as the deliberate 'governance of' socio-technical systems, we relate to planning as the project of changing the social mechanisms that govern social action. Sustainable transition understood as normative planning going beyond institutionalised patterns of agency becomes a question of addressing governance structures of socio-technical systems with the aim of changing the pattern of governance activities. Sustainable transition, in this understanding, becomes a project of 'meta-governance', the development of strategies and capabilities to influence and shape governance structures of socio-technical systems. From a change-oriented planning perspective, it is, therefore, important to identify how to influence the activities of governance. It is a metagovernance perspective, departing from the understanding that the important role of deliberate governance (planning) lies in the shaping the fields of actions thereby conducting the conduct of governance activities (Dean 1999: 10–13) and influencing the way that regimes form cognitions, norms and practices that structure the actions of private and public actors and frame processes of innovation and diffusion of technology – and shape technological paths.

Agency and planning in the fields of socio-technical systems always have to be judged in terms of their restructuring effect on the governance structure. Taking this broader 'restructuring approach' also implies a contextualisation of the transition processes. Governance structures are, as stated initially, subject to competing contemporary modernisation agendas of restructuring; all influencing governance activities of socio-technical systems.

'Governance structure' may be seen as inherent in the concept of governance, however, we find it important to emphasis that we have such a governance structure, formed by all the constituent components of the socio-technical system within which governance processes take place. Governance activities have to be seen as structured by such mechanisms as systemic interdependencies and configurations of institutions and actors, structures with a high degree of rigidity and resistance to changes.

The Dynamics and Transition of Socio-technical Systems: The Multi-level Model

The normal practices of socio-technical regimes in relation to sustainability are incremental improvements and optimisation. Obtaining sustainable modes of production, however, takes more radical changes in terms of system innovation (Weaver et al. 2000, Vellinga 2001) or transitions changing the structures and rationalities of the socio-technical regimes (Loorbach 2007). The challenge concerns how shifts in norms and practices of these regimes (in governance structures) can be addressed with the aim of shaping (mediate the change of) the governance activities of the regimes. Addressing this problem, we have to take into account the complexity, ambiguity and uncertainty related to sustainable transitions of socio-technical systems.

On a general level, changes of systems are related to changes in the elements of the system, in their relations within the system or with the environment of the system. Taking an evolutionary approach, development is understood as processes of variation and selection, making deliberate changes a question of inducing variety and shaping selection environments (Kemp et al. 1998). Building on this understanding a 'multi-level model' has been introduced (Rip and Kemp 1998, Geels 2002, 2004a, 2004b). In this model changes are perceived as a multi-level process; an interplay of the development of *niches, regimes and socio-technical landscape.* Figure 5.1 gives a graphic presentation of how socio-technical regimes (systems) are changed driven by emerging niche technologies, internal processes of the regime, and changes in the social political landscape – resulting in reconfigured socio-technical regimes.

Landscapes include such elements as material infrastructure, the macro-economy, political culture and coalitions, social values, belief systems and paradigms, demography and nature. Thus landscapes are external factors of importance for the socio-technical regime. Landscape changes can put pressure on existing regimes

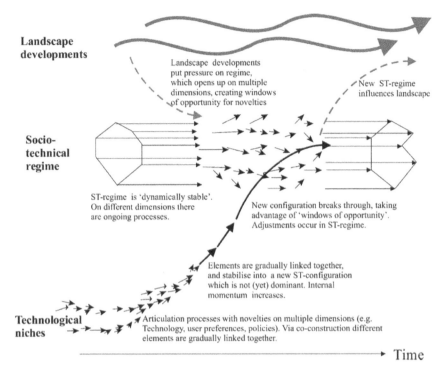

Landscape
developments

Landscape developments
put pressure on regime,
which opens up on multiple
dimensions, creating windows
of opportunity for novelties

New ST-regime
influences landscape

Socio-
technical
regime

ST-regime is 'dynamically stable'.
On different dimensions there
are ongoing processes.

New configuration breaks through, taking
advantage of 'windows of opportunity'.
Adjustments occur in ST-regime.

Elements are gradually linked together,
and stabilise into a new ST-configuration
which is not (yet) dominant. Internal
momentum increases.

Technological
niches

Articulation processes with novelties on multiple dimensions (e.g.
Technology, user preferences, policies). Via co-construction different
elements are gradually linked together.

Time

Figure 5.1   A dynamic multi-level perspective on system innovations
Source: (Geels 2004a, 915).

(Geels 2002, 2004a) or may open windows of new opportunities (e.g. changes such as those resulting from a global economic crisis or new climate agenda).

Niches are local domains, where non-standard technologies and new learning processes materialise. They are domains where (potentially) new elements in terms of technology, new actors, practices, institutional arrangements, etc. emerge. Niches may be protected from the normal markets and they often, as places of technology development, deviate from some of the rules in the dominant regime (Geels 2002, 2004a) or even are related to actor networks (social entrepreneurs) with alternative values. As they develop, they may cluster with other niche technology to form more coherent alternatives to dominant technologies, and they may develop their own 'niche regime' stabilising development paths and enabling attraction of resources and actors (Loorbach and Rotmans 2009: 116).

Conceived within this framework, the change of socio-technical *regimes* for sustainable development can be related both to changes in the landscape (e.g. changes in normative public values, in price structures and so on), tensions within the regime and/or to the development of niches. Processes of change can take many routes; they can be driven through bottom-up processes based on social movements (Smith 2003) or as a deliberate process of transition management (Kemp et al. 1998, Kemp and Rotmans 2001).

We have a dynamic relation of structuration (Giddens 1976) between the levels. Socio-political landscapes are the media structuring the development of regimes but are simultaneously transformed as the outcome of developments in the regimes. At the same time, we have a similar interplay of niches and regimes.

The integration of wind energy into the socio-technical system of heat and power in Denmark is an illustrative case. The modern development of wind energy in Denmark started as a niche technology carried by social movements and manufactures of agricultural machineries with over-capacity. After a long process of experimentation and selection, the design stabilised on the stall-regulated, upwind, three-bladed turbine. During the 1980s the technology matured. Actors, networks and institutions developed and the industry transformed from grass roots-based production to an industrial network. During the 1990s, due to changed institutions (in particular improved access to the grid), there was a rapid growth in the number of windmills. However, in 2001 the expansion levelled out due to a redefinition of the institutional conditions. Wind power had become an integrated part of the dominant socio-technical system of heat and power, but due to uncertainty about the external costs of energy produced on fossil fuel and uncertainty about governmental support to wind energy, further expansion in Denmark became uncertain until a broad coalition in the Danish Parliament, in 2008, decided on a new plan for the development of wind power as an important element in the supply system.

Now we have a highly developed and mature wind technology niche regime developing in interaction with the energy regime. There is a shift at the energy regime level: the development of flexible energy systems capable of integrating wind energy as a major energy source has become a main goal in European energy programmes.

Deliberate Governance of Socio-technical Systems

Following the multi-level model, the planning field of deliberate sustainable transition includes the processes developing in and between changes in landscapes, socio-technical regimes and niches. This relates to a concept of planning not focusing on identifying a particular position from where 'somebody' can govern future development. Planning is perceived as an activity undertaken in many places in society, conducted by many actors, and reflexive in relation to goals and change strategies.

Socio-technical regimes are complex systems that defy blueprint steering also in relation to a deliberate transition for sustainability, both due to the character of the system and due to the character of the challenge of sustainability. According to Voß and Kemp (2006) deliberate transformation will face fundamental problems related to the complex and heterogeneous nature of the system, problems of goal formulation as well as strategy implementation.

It is impossible to obtain *a complete analytic understanding of socio-technical systems*. The deliberate transformation of socio-technical systems and their interplay with ecological systems requires knowledge about the very heterogonous elements of these systems, for example technological artefacts, networks, chemicals in the atmosphere, regulatory institutions. Knowledge about how these elements interact and develop is well beyond conventional disciplinary science. And as the full complexity may be beyond what scientific methods can handle, a more synthetic and practical form of knowledge gained from practical experience by the societal actors is needed. Thus, transition strategies must include knowledge production that transcends both the boundaries between scientific disciplines and the boundary between science and society/social practice experiences.

Furthermore, the complexity and heterogeneity of the socio-technical systems and their interplay with ecological systems makes change a complex and uncertain process. The sheer number of interconnected processes and the non-linear character of many social and ecological systems preclude prediction of the long-term effects of interventions. Therefore the second requirement is that strategies must be experimental and adaptive to allow for error and learning. Thirdly, and paradoxically, the path-dependent character of the development of socio-technical systems makes careful anticipation and assessment of potential development paths an important element of any strategy process (Voß and Kemp 2006: 10–14).

In a planning perspective, a main implication is that the knowledge of social-technical systems has to be acquired in reflexive practices. This implies a need for a 'mode 2' science in which knowledge production *is socially distributed, application-oriented, trans-disciplinary, and subject to multiple accountabilities* (Nowotny et al. 2003: 181).

The second problem relates to *goal formulation*. It can be argued that in principle sustainability is a functional requirement for societal development, a claim for a development that can be sustained without eroding its own foundation. If the development of social and ecological systems could be predicted and the resilience and threshold values of these systems were well known, it would be a task for science to demarcate sustainable and unsustainable development paths. But social and ecological development is difficult to predict, and knowledge about resilience and threshold values is only partial and often disputed. Therefore what science can deliver are more or less well-founded assessments of the probable consequences of different types of impacts and developments. Therefore risk assessment is crucial, determining which impacts and developments are acceptable or desirable. And as risk assessment is based on values, the assessment of what is sustainable is in the end a very value-loaded exercise. And values vary amongst people and change over time. Therefore the assessment must be participatory – as everybody's values must be included – and ongoing, as values change (Voß and Kemp 2006: 14–16).

Again, it implies an understanding of the research process, where 'it can no longer be characterized as an "objective" investigation of the natural (or social) world, or as a cool and reductionist interrogation of arbitrarily defined "others". Instead, it has become a dialogic process, an intense (and perhaps endless) "conversation" between research actors and research subjects' (Nowotny et al. 2003: 187).

Finally, we face the problems of *strategy implementation*. The development of socio-technical systems is shaped in the daily interaction between consumers, producers, public officials and policy makers, amongst other actors. Thus the capacity to influence the development of socio-technological systems is distributed amongst a broad range of actors each having their views, interests and resources, and directed change of the systems requires coordinated action amongst all the actors (Voß and Kemp 2006: 16–17).

Opposed to a traditional planning approach, Voß and Kemp (2006: 18) assert that efforts of deliberate sustainable transition of socio-technical systems have to be understood as a process of reflexive governance involving an iterative, participatory and reflexive approach to system analysis, goal formulation and strategy implementation. And, as indicated above, a need of knowledge of transition which is acquired in reflexive practice.

Transition becomes a question of addressing and changing governance activities of socio-technical systems. We have a tension between this understanding of socio-technical systems developing governance through social practices and the intention of acting for deliberate sustainable transition. Although processes of socio-technical systems are understood as distributed, reflexive governance installs a privileged external position of metagovernance (deliberate change of governance structures in socio-technical systems). This tension becomes even more evident when we address attempts to develop transition management strategies.

Transition of Socio-technical Systems: Strategies and Experiences

The question of how socio-technical systems can be deliberately transformed taking into account complexity, path dependency, goal ambiguity and outcome uncertainty, has been subject to both scientific scrutiny and political consideration. The problem is how the general request for a sustainable reflexive modernisation process (Beck et al. 2003: 6) or reflexive governance can be converted into operational policy, programmes and means for transition of unsustainable practices withheld by our present socio-technical systems. To unfold this discussion we are going to address two cases of 'transition practices', namely the Dutch research and policy programme of 'transition management' (Kemp and Rotmans 2001, Kemp and Loorbach 2006, Loorbach 2007) and the EU's climate strategy. Initially, we introduce the issue of how 'transition management' has faced the governance problems outlined above.

Transition Management

The point of departure of transition management is that sustainable development calls for long-term, radical and structural changes of the socio-technical system and that the uncertainty related to the complexity of the system and the ambiguity of the goal of sustainability rule out instrumental blueprint planning approaches based on direct control, fixed goals and predictability. Departing from this understanding, the proponents of transition management have called for a planning and policy approach which is reflexive in terms of visions and goals; is based on goal-oriented incrementalism; and which is multi-actor and multi-level oriented. On an operational level the transition management programme has established a set of guiding principles (Kemp and Loorbach 2006, Loorbach 2007).

There has to be an orientation to long-term transition flexible goals (based on sustainable visions) as a framework of the development of short- and mid-term learning and innovation programmes related to these goals. This is a process of anticipation and adaptation (Loorbach 2007: 82). The starting point is a social process among the involved actors of the socio-technical system where collective sustainable visions are articulated and turned into long-term transition goals. Based on these visions and goals, including a sense of urgency, short-term agendas and action programmes are shaped in a bottom-up process, forming a transition of the socio-technical system. The process in total has to be adaptive, in terms of a continuous, cyclical and reflexive process of social learning related to programmes and experimentation (see below), which is turned into reassessment and adjustment of goals and programmes.

There has to be a systemic-oriented thinking in terms of a multi-domain, a multi-level and a multi-actor approach. This is based on the understanding of socio-technical systems presented above, stressing the need for concurrent work with cultural, social, economic and technical aspects (the need of aligning different political domains in the change processes) of the socio-technical system and an understanding of changes of socio-technical systems as involving change processes at all levels and dependent on the distributed capacity in the system.

There should be a focus on system innovation and experimentation/learning (create variety based on visions). This is an attempt to obtain substantial and radical changes of structures and rationalities of socio-technical systems. An experimental and learning approach is important to ensure a variety in options and avoid premature lock-in to specific paths of technology and development.

There should be an awareness of the need of an opening of policy processes to reduce the dominance of vested interests and escape lock-ins. Escaping lock-ins also involves the formation of new actor constellations, both to break path dependence related to vested interest and to involve new visions, values and resources related to new actors such as grass roots movements and entrepreneurial front-runners. Part of this can be the reconfiguration of institutional arrangements favouring existing technologies and incumbent players.

Finally, the transition process is perceived as social processes based on participation and interaction between stakeholders. Transition is going to be rooted in the shaping of collective goals and joint decision making, where agendas and programmes are developed in interaction with stakeholder groups. A vital part of transition management is organising and facilitating interaction, but still leaves the formation of vision, goals, agendas and programmes to the interaction between the involved stakeholders. Transition management is perceived as 'a forward-looking, adaptive, multi-actor governance aimed at long-term transformation processes that offer sustainability benefits' (Kemp and Loorbach 2006: 103).

A central instrument in turning transition management into operational policy and planning is the formation of multi-actor networks (transition arenas) in relation to specific sectors (innovation systems) and transition goals, with the aim of establishing new rationalities and capacities of the innovation system. In general terms, the transition arena is conceived 'as a new institution for interaction [that] can be considered a meta-instrument for transition management and facilitates interaction, knowledge exchange and learning between the actors' (Kemp and Loorbach 2006: 111). The major objectives of transition arenas are to define problems, establish transition visions and transition goals, and to create public support and broadening the coalition.

The arena (or activity cluster) is intended as the basis of a reflexive governance process. In a cyclic way it is supposed to move from stages of organising multi-actor networks, developing (negotiating) sustainability visions and transition agendas, mobilising actors, executing projects and experiments, and evaluating, monitoring and learning (see Figure 5.2). It should condition the process of bringing forward

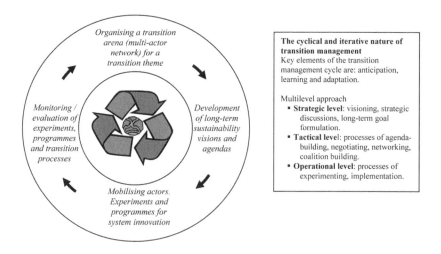

Figure 5.2   Transition arenas and the cyclical process of transition management

alternative (niche) technologies, e.g. facilitate and modulate the generation of a variety of technology options and selection processes. Establishing such arenas is a crucial point – and as the Dutch case of transition of energy systems will show, a contested point.

Transition of the Dutch Energy System

Since 2001 transition management has been part of official policy in the Netherlands. The approach was introduced in the National Environmental Policy Plan 4, NEP-4 (Ministry of Housing, Spatial Planning and the Environment (VROM) 2001). When the plan was prepared there was an openness amongst policy makers towards new approaches and the idea of 'transition management' fitted the needs of a coalition of policy makers well. The approach was developed in a close cooperation between policy makers and scientists and formed the cornerstone of NEP-4 (Smith and Kern 2007, Loorbach 2007: 159–67). Transition management has been explored within energy, agriculture, biodiversity and mobility in the Netherlands and the policy concept has, since its introduction in 2001, steadily gained momentum (Loorbach 2007: 284–5).

An early attempt to implement the strategy was the energy transition project 'EnergieTransitie'. The project was developed in cooperation between the Ministry of Economic Affairs and a group of scientists. In the early phase, three quality criteria were formulated for the future energy system: security of supply, economic efficiency and sustainability. Based on different scenarios for the development of the Dutch energy supply system, four main transition routes were identified: (1) new and efficient and green gas; (2) modernisation of energy chains; (3) Biomass International; and (4) Sustainable Rijnmond (an industrialised and urbanised region in the Netherlands).

Transition arenas were established for each of the four transition roads, more specific visions for the future were developed and altogether 80 ideas for transition experiments were collected. In 2005, a so-called Taskforce Energy Transition was set up with 15 high-level representatives from science, business, NGOs and the government led by the CEO of Shell Netherlands. Its task was to reflect upon the overall process of energy transition and to define a shared direction. Furthermore, an 'Interdepartmental Projectdirectorate Energytransition' with officials from the involved ministries was set up to reflect upon the process in order to govern the process better and to develop innovative policies (see Loorbach 2007, Smith and Kern 2007, Ministry of Economic Affairs 2004).

Assessed as a policy process, the energy transition process has clearly been a success for the Ministry of Economic Affairs. What started as a policy experiment around the year 2000, with few actors and a minor budget, was in 2007 a well-consolidated programme with several hundred actors involved and more than 70 ongoing transition experiments (Loorbach 2007: 267–9).

Assessed as an example of transition management, the Energy Transition Project has deviated from the initial planning visions described above. The first phases of the process in particular took another route. Instead of establishing a transition arena where selected innovators and front-runners could develop shared visions for the future, the process was more focused on establishing a consensus about visions in a stakeholder network of incumbent actors (Loorbach 2007: 256). There may be several reasons for this. One is that the transition management approach was not fully developed when the energy transition project started (Loorbach 2007); another might be the strong interests of a number of the central actors (Smith and Kern 2007).

Whatever the reasons, the consequence was that the focus narrowed to sustainable business opportunities and that the behavioural, institutional, structural and cultural changes necessary for a transition of the energy system were more or less ignored (Loorbach 2007, Smith and Kern 2007). Thus, assessed as an attempt to initialise a transition of the Dutch energy system, the Energy Transition Project came short. Even though new actors have become involved, new coalitions established, public debate stimulated, the bottom line is that the Energy Transition Project can best be seen as an advanced energy technology innovation programme that can contribute to optimising the existing dominant energy system. Judged by the track record of the project, it seems unlikely that the Energy Transition Project will contribute significantly to a transformation of the Dutch Energy system.

To some extent these experiences can be seen as first mover problems; however we find that they expose substantial problems of the deliberate transition of socio-technical systems. Jänicke (2007: 8) has asserted that changing the dominant socio-technical systems of modern society 'is not going to be easy', and he has warned that the advanced management of innovation processes does not do the job. Transition scholars have to some degree admitted this by saying that 'so far little attention has been directed towards issues of power, institutions and leadership' (Loorbach 2007: 294). Given the way transition processes are conceptualised in transition management, it is not surprising that power and institutions evade attention. Even though the importance of institutions for the direction of innovation and power for the stability of socio-technical regimes are explicitly acknowledged in the analysis of the dynamic, but path-dependent, development of socio-technical systems, these crucial insights are omitted in the conceptualisation of transition management. The problem surfaces clearly in the thinking about how 'transition visions' and 'transition arenas' are to be established.

The establishment of transition visions is conceptualised as a top-down process in which an 'inspiring, imaginative and innovative vision is translated into transition images at a sub-system or thematic level' (Kemp and Loorbach 2006: 113). Although much emphasis is put on the openness of the visions and the participatory and iterative character of the process of formulating them, one vision is supposed to be guiding for activities at all levels.

Transition arenas are established by selecting visionaries, forerunners, open-minded persons. The selection is seen as of 'vital importance' (Kemp and Loorbach 2006: 112), but there is no guidance on who should select them or on the selection process. However, inevitably, somebody else in a top-down process must select the participants. Moreover, in an open democratic society there will be a tendency towards a strong representation of the incumbent actors in the transition arena.

The probable result of this conceptualisation is exactly what we have seen in the Dutch energy transition example: transition arenas dominated by incumbent actors and a transition vision formulated on the premises of the existing problematic socio-technical system – and influences from parallel competing agendas. Innovation is seen as technical innovations within the given framework developing 'sustainable business options' and the likely result is incremental innovation and not system transformation.

This could be reformulated in the following way: the consequence of the way the transition management is conceptualised is that important cognitive and normative institutions are established in a top-down process that generally will give preferences to the institutions of the incumbent actors. This implies that only innovations that are compatible with the main institutions of the dominant and problematic system can be part of a transition project. Thus, the likely result is that what should have been transition management turns out to be sustainable business innovation.

Low Carbon Energy Systems: In a Transition Perspective

Transition to a low carbon energy system has been identified as one of the EU's main goals (European Council 2007), setting an agenda of a long-term radical shift in the existing system. Although it is not conceptualised as a transition management project, it stipulates a shift in a mature socio-technical system. The examination of transition initiatives taken can contribute to our understanding of the challenge of deliberate governance of transition of socio-technical systems.

The development of the European energy system has been dominated by a market approach. Energy production has been privatised and market institutions have been set up to shape energy markets. Climate change and agendas of securing energy supply and independency (changes in the socio-political landscape), however, have introduced a need for transforming the established socio-technical energy system. This has spurred the introduction of 'The energy and climate change package' (European Council 2007, European Commission 2008), stating a 20 per cent reduction of CO_2 (compared to 1990) and 20 per cent renewable energy in 2020 as its main goals.

In its report on the 'Strategic Energy Technology Plan' (EU Commission, 2007), the European Commission acknowledged that public intervention was needed if the new technologies and a transition to low carbon energy system should be accomplished. Deliberate transition is accepted as necessary, but both the strategies and subsequent climate policy action programmes show high path dependency and an inability to break with the established governance structure in the sector. These problems have been exposed during the process and in the policy programme outlined in the energy and climate change package. Three problems can be highlighted:

1. The economic discourse dominates. The goal of accelerating the development and deployment of cost-effective low carbon technologies pre-empt more radical and visionary transition strategies.
2. The established governance structures are preserved. The CO_2 Emission Trading System and the Cleaner Development Mechanism/carbon credits (the package allows up to three-quarters of the CO_2 reduction to be accomplished outside Europe) are maintained as the main instruments. Radical actions, such as a substantial reduction of CO_2 quotas and the introduction of CO_2 thresholds for new power plants, are not included due to the interests of incumbent market players. In general, a strong belief in the CO_2 market is asserted.
3. The formation of new technology paths is biased by incumbent players. A significant example is the position carbon capture and storage (CCS) technologies has acquired as a main element in the implementation of 'The Energy and Climate package'. As a CO_2 abatement technology CCS is at a very early stage with no full-scale plant experience. It demands 20–30 per cent higher energy consumption in power production, even if problems of scaling up the process technologies are solved. Despite these drawbacks, CCS (together with nuclear power) gets the main part of development funding within the programme (€10 billion to 10–12 CCS demonstration plants 2013–2020, financed by the sale of CO_2 quotas). This outcome of the policy process can be ascribed a combination of strong interests of coal and oil sector as well as of coal-dependent member states and an economic assessment in which CCS is identified as the most cost-effective way of obtaining CO_2 reduction (v. Alphen et al. 2009).

In a transition perspective, the EU low carbon system programme faces problems resembling the Dutch case. In particular, the definition of CCS as a main path in the EU's programme of low carbon energy system, illustrates a number of problems of undertaking deliberate transition of mature systems:

1. Incumbent actors succeed to introduce their interpretation of low carbon system and succeed in defining selection criteria favouring their technology choices. New actors are not given access to the process of shaping the programme.
2. We have a systemic selectivity by the dominant regime. CCS represents a technology which fits well with the established energy system. It can be introduced as an add-on technology without introducing changes in the energy system as such.
3. We have a selective and biased interpretation and perception of the climate and energy programme, reducing the challenge to providing cost-effective low-carbon technologies (leaving a broader request for sustainability unaddressed). Within this understanding, the introduction of CCS makes perfect sense.

4. We experience an inability (or lack priority) to define and form new transition arenas. The institutional framework and actor relations established to deploy an energy market give a weak platform for staging transition processes. The established governance structure is left unchanged.
5. The transition perspective is overruled by other agendas, such as preserving the vested interests of the private sector – conflicts of paths are resolved within an economic discourse, not a sustainability discourse.

The development of a dynamic and flexible energy system with a high capacity to integrate VE-energy (in particular wind power) has been curbed by CCS programmes supported by the vested interests of the established system.

Deliberate Governance of Transition of Complex Socio-technical Systems

The experiences with ecological modernisation, Dutch transition management and the EU strategy for the transition to low carbon energy systems offer valuable inspiration for the deliberate governance of transition of complex socio-technical systems, and give substance to reflections about the difficulties and dilemmas involved in the governance of transition processes. Governance has to cope with technological and institutional path dependencies, incumbent actors with vested interests, lack of well-defined goals and knowledge, societal priorities, risks and complexity.

An important insight from ecological modernisation is that it is possible to establish institutional learning processes around environmentally friendly technologies. Often, though, the technologies have been stand alone technologies and the scope of improvement has been limited because they have been embedded in unsustainable socio-technical systems. The EU strategy for the transition to a low carbon emission system shows that even when new discourses evolve that change the landscape, there is a risk that incumbent actors capture the right to define the problems and solutions thereby sustaining the dominant regime. Even more important, the case demonstrates a need to define 'transition management' much more broadly. 'Transition management' also has to be concerned with counteracting other transitional restructuring agendas and activities – a transition programme has to be constantly reflexive concerning how governance structures are redefined.

The transition management programme has forwarded the most ambitious and explicit strategies for changing socio-technical systems. A core problem is, however, that transition management has not acknowledged the full consequences of the understanding of innovation as taking place within a system composed of different cognitive and normative institutions. The way the multi-level model is used as the conceptual framework for transition management tends to conceal all the conflicts in cultures and norms around innovations and technology between the different niches and between the niches and the dominant system. Thus, it misses the variation in cultures, norms and interpretations about what needs should be fulfilled and in which ways, and conflicts over visions of the future are a prerequisite for variation in innovation. Or put differently, variation in cognitive and normative institutions is a prerequisite for variation in innovation.

If radical innovations are to take place, there has to be room for the development of niches in which the conception of what a sustainable future looks like, which needs should be fulfilled, how, which technologies are promising, etc. can be discussed. There is a need for experiments, including with technologies, which are incompatible with, and challenge, the dominant cognitive and normative institutions. That is not very likely to happen if there is *one* vision of transition and *one* transition arena (for a parallel discussion of changes of regimes, see Berkhout, Smith and Stirling 2004).

Metagovernance has to address the fact that cognitive and normative institutions are decisive for innovative processes. To facilitate system transition, room for different and conflicting interpretations of what a sustainable future could look like must be accepted and supported during the process. There is a need for the '*agorae*', as places of controversies/agony and of problem-generating and problem-solving processes involving competing positions and interests (Nowotny 2003: 191, Pløger 2004). This is necessary to get enough variation into the innovation process and improve the likelihood of radical innovations. This implies that governance actively should pursue the formulation of different visions for the future and should establish not one, but several transition arenas.

However, doing so would enhance the dilemmas of steering transition in at least two ways. Firstly, the transition process itself would become more agonistic because there would be differences between the various niches' understanding of the goal of the innovation process and conflicting demands about how the common regulative institutions should look in the future. Thus, there inevitably would be an ongoing debate between the different niches regarding what actually is sustainable and how the regulative institutions ought to develop. Secondly, the political conflicts around the transition programme would intensify. Instead of supporting developments that are more or less compatible with the existing incumbent actors, transition projects would now support both compatible niches and conflicting niches that if turned into the dominant socio-technical systems would eradicate most of the existing incumbent actors. That the incumbent actors of course will resist the best they can.

Expanding Transition Strategies

Taking a more conflict-oriented approach to the process and taking the systemic nature of innovation and the need for variation seriously makes the governance process more complicated.

The consequence of acknowledging the systemic nature of innovation and the need for variation is that each niche must be understood as a system on its own with its own needs to be fulfilled if it is to develop. Or put differently, if a niche is to develop a number of functions have to be fulfilled. Some of the important functions are the creation and diffusion of knowledge, guidance of the direction of search among users and providers of technology, the formation of markets, legitimating and resource mobilisation such as capital and competences (Jacobsson and Bergek 2004, Bergek et al. 2005, Loorbach and Rotmans 2009). Doing so both requires an appropriate governance structure within the niche but also an appropriate external governance structure.

Based on these arguments we maintain that the following are the main demands to the governance of transition processes:

1. *Securing variation and accepting conflict.* Firstly it is necessary to replace the idea of *one* transition arena with the idea of a transition *field* with a multitude of transition arenas each related to a number of niches. As argued above, establishing only one transition arena will tend to favour the incumbent actors and their understandings and strategies, which will eliminate the possibilities for radical innovations and most likely only lead to system optimisation. We would argue that it is much more productive to aim at establishing different transition arenas where different groups of actors can develop their operational meaning of sustainability and attempt developing technologies, cultures, norms, practices, lifestyles that comply with this understanding (see Smith 2003, Nielsen and Nielsen 2003, Heinberg 1997).

2. *Taking the politics* of *transitions seriously.* Genuine system transformations are highly controversial and political processes because in many cases they will challenge incumbent actors and the lifestyles of many people. So far, the focus of the transitions scholars has mainly been on a policy *for* transitions, but we also need a politics *of* transitions (Kern and Smith 2008, Shove and Walker 2007). Securing stable long-term support for transitions is not easy as the incumbent actors in the big socio-technical systems of modern society will normally resist transition and are normally very powerful. Whereas the actors in the niches of the technologies of tomorrow are normally weak and poorly organised. To counterbalance the incumbent powerful actors, strong support for the necessity of system transition towards sustainability in the general public is indispensable. Only such support would make it politically possible to support technology niches that are incompatible with the dominant socio-technical system, and in the long run challenge the position of the incumbent actors.

3. *Complicating and differentiating the governance approach.* Securing variation, accepting conflict and taking politics seriously have profound consequences for the governance structure that has to be differentiated both horizontally and vertically.

   a. Horizontally there is a major distinction between the establishment of and subsidies to transition arenas, and the governance within each of the transition arenas and the niches connected to them. We would argue that it is necessary to establish a politically dominated coordinating forum. The task of such a coordinating forum would be to establish transition arenas, ensure the necessary economic and institutional support to the different competing and conflicting transition arenas, and prioritise and coordinate amongst them, and in general, it would be the arena for the politics *of* transitions.

b. Vertically there would be a differentiation between the governance approach of the different transition arenas. Each transition arena would develop its own governance approach together with the niches connected to it. Such a governance approach with a multitude of transition arenas and a coordinating forum would strengthen variance in the innovation of technologies, practices and lifestyles, and would escape, to some degree, the dilemma of supporting front-runners versus incumbent actors, would make it easier to maintain broad support to a transition programme.

Governance of Change?

We have analysed the governance structures of socio-technical systems in order to identify possible roads for the governance of sustainable transitions. It is a planning problem based on a paradox. For the last 20 years, the concept of sustainability has (at least in some periods) been a steering point for planning in the public domain and policy programmes of sustainability have been developed. At the same time, one could argue that the main socio-technical systems such as energy production, transportation, food production and the building industry have developed (path dependently) in a non sustainable direction.

Our analysis of ongoing transition strategies does not provide a simple answer. Transition management as a programme has developed some answers to the dilemmas of deliberate governance of socio-technical systems, but even this programme is a project in doubt of its own capability – stuck in a conflict in addressing an unmanageable complex system with distributed governance activities beyond the reach of any planner and a simultaneous imperative to act against unsustainable systems of production and consumption.

An important implication of the governance perspective presented here is that transition cannot be isolated to 'transition projects'. Governance structures of socio-technical systems are constantly being restructured driven by competing agendas and distributed agency and power. Sustainable transition as a normatively driven project becomes a project of intervening in these projects. Deliberate governance for sustainable transition has to penetrate all levels of governing and planning, addressing processes of niches, regimes and landscapes and their interaction, and to this end it has to be founded on commitments to politics for sustainable transition. It demands a constant reflexivity about how any action and agency restructures governance structures of socio-technical systems, how power and resources are redistributed, how new arguments and rationalities are given access, how new configurations of technologies, actors and institutions are formed not only by deliberate transition actions but in every action.

There is a need for taking deliberative governance of transition processes in complex socio-technical systems seriously. Governance in the upcoming socio-technical system has to focus on establishing learning communities in order to develop technologies not yet seen or to mature infant technologies. Networks of actors from knowledge institutions, business institutions and regulatory institutions have to be established and it is important that these networks are open to the participation of social and environmental entrepreneurs.

'Pick the winner' strategies have to be avoided. Therefore it is important to support niches in the same area with conflicting coalitions of actors, perceptions and strategies. Taking the energy system as an example, it is not possible in advance to decide whether wind turbines, biofuel or biomass for district heating is the cornerstone in the future energy system. Actually, it is more likely that a sustainable energy system combines different types of energy sources.

Planning has an important role to play in order to establish the institutional framing for learning communities, standards and markets and in order to develop the governance structure in the socio-technical systems. Sustainable technologies will often find it difficult to compete with mature systems. Therefore, standards, public procurement, market policies and tax policies have an important role to play. The setting of specific goals and policies has to be flexible.

Finally, we will emphasise that sustainable transition is a political project; the formation of politics for sustainable transition is a prerequisite for deliberate transition. In a Danish context, public policies in the 1980s and 1990s combined with social and environmental movements and entrepreneurs, led transitional processes in the direction of more sustainable socio-technological systems in the areas of food, energy and clothing. Shift in policies, following a new Danish Government in 2001 cancelled this support and instead supported a 'climate change denial project' which stalled these transition processes. The definition and empowerment of sustainable transition policies is a first prerequisite, the necessary platform of deliberate governance of transition.

References

van Alphen, K., van Ruijven, J., Kasa, S., Hekkert, M.P. and Turkenburg, W.C. 2009. The Performance of the Norwegian Carbon Dioxide, Capture and Storage Innovation System, *Energy Policy*, 37(1), 43–55.

Arthur, W.B. 1990. Positive Feedbacks in the Economy, *Scientific American*, 262(2), 92–9.

Beck, U. 1994. Self-dissolution and Self-endangerment of Industrial Society: What Does it Mean?, in *Reflexive Modernization: Politics, Tradition and Aesthetics in the Modern Social Order*, edited by U. Beck, A. Giddens and S. Lash. Cambridge: Polity Press, 174–83.

Beck, U. 2006. Reflexive Governance: Politics in the Global Risk Society, in *Reflexive Governance for Sustainable Development*, edited by J-P. Voß, D. Bauknecht and R. Kempt. Cheltenham: Edward Elgar.

Bergek, A., Jacobsson, S., Carlsson, B., Lindmark, S. and Rickne, A. 2005. *Analyzing the Dynamics and Functionality of Sectoral Innovation Systems: A Manual*. Paper presented at The DRUID Tenth Anniversary Summer Conference 2005 on dynamics of industry and innovation: Organizations, networks and systems, Copenhagen, 27–29 June 2005.

Berkhout, F., Smith, A. and Stirling, A. 2004. Socio-technological Regimes and Transition Contexts, in *System Innovation and the Transition to Sustainability*, edited by B. Elzen, F.W. Geels and K. Green. Cheltenham: Edward Elgar, 48–75.

Callon, M. 1991. Techno-economic Networks and Irreversibility, in *A Sociology of Monsters? Essays on Power, Technology and Domination*, edited by J. Law. London: Routledge.

David, P.A. 1985. Clio and the Economics of QWERTY, *The American Economic Review*, 75(2), 332–37.

Dean, M. 1999. *Governmentality, Power and Rule in Modern Society*. London: Sage.

Dosi, G. 1982. Technological Paradigms and Technological Trajectories, *Research Policy*, 11(3), 147–62.

European Commission. 2006b. *A European Strategy for Sustainable, Competitive and Secure Energy*. Green Paper. COM(2006) 105 final.

European Commission. 2007a. *Towards a European Strategic Energy Technology Plan*, 10.1.2007, COM(2006) 847 final.

European Commission. 2007b. *A European Strategic Energy Technology Plan (SET-plan) – Towards a Low Carbon Future*, 22.11.2007, COM(2007) 723 final.

European Commission. 2008a. *20 20 by 2020. Europe's Climate Change Opportunity*, 23.1.2008 (2008) 30.

European Commission. 2008b. *Impact Assessment*. Directive on the Geological Storage of Carbon Dioxide, 23.1.2008, SEC (2008) 55.

European Council 2007. *Presidency Conclusions of the Brussels European Council*, 8/9 March 2007.

European Council. 2008. *Decision _OTE*, General Secretariat of the Council to: Delegations Subject: Energy and Climate Change: Elements of the Final Compromise, 12.12.2008 17215/08.

Geels, F.W. 2002. Technological Transition as Evolutionary Reconfiguration Processes: A Multi-level Perspective and a Case Study, *Research Policy*, 31(8/9), 1257–74.

Geels, F.W. 2004a. From Sectoral Systems of Innovation to Socio-technical Systems: Insights about Dynamics and Change from Sociology and Institutional Theory, *Research Policy*, 33(6–7), 897–920.

Geels, F.W. 2004b. Understanding System Innovations: A Critical Literature Review and a Conceptual Synthesis, in *System Innovation and the Transition to Sustainability: Theory, Evidence and Policy*, edited by B. Elzen, F. Geels, and K. Green. Cheltenham: Edward Elgar.

Geels, F.W. 2005. The Dynamics of Transitions in Socio-technical Systems: A Multi-level Analysis of the Transition Pathway from Horse-drawn Carriages to Automobiles (1860–1930), *Technology Analysis and Strategic Management*, 17(4), 445–76.

Giddens, A. 1976. *New Rules of Sociological Method: A Positive Critique of Interpretative Sociologies*. London: Hutchinson.

Grin, J. 2009. Understanding Transition from a Governance Perspective, in *Transition to Sustainable Development: New Directions in The Study of Long Term Transformative Change*, edited by J. Grin, J. Rotmans and J. Schot. New York and London: Routledge.

Heinberg, C.H. 1997. Skal tumberne redde verden – når nu de kloge ikke vil? in *Livet i Drivhuset*, edited by I. Geertsen. Copenhagen: Det Økologiske Råd.

Hughes, T.P. 1986. The Seamless Web: Technology, Science, etcetera, etcetera, *Social Studies of Science*, 16(2), 281–92.

Hughes, T.P. 1993. *Networks of Power: Electrification in Western Society, 1880– 1930.* Baltimore: John Hopkins University Press.

Jacobsson, S. and Bergek, A. 2004. Transforming the Energy Sector: The Evolution of Technological Systems in Renewable Energy Technology, *Industrial and Corporate Change*, 13(5), 815–49.

Jänicke, M. 2007. Ecological Modernisation: New Perspectives, *Journal of Cleaner Production*, 16(5), 557–65.

Kemp, R. and Loorbach, D. 2006. Transition Management: A Reflexive Governance Approach, in *Reflexive Governance for Sustainable Development*, edited by J-P. Voß et al. Cheltenham: Edward Elgar.

Kemp, R. and Rotmans, J. 2001. *The Management of the Co-Evolution of Technical, Environmental and Social Systems.* International Conference towards Environmental Innovation Systems, Garmisch-Partenkirchen, September 2001.

Kemp, R., Schot, J. and Hoogma, R. 1998. Regime Shifts to Sustainability Through Processes of Niche Formation: The Approach of Strategic Niche Management, *Technology Analysis and Strategic Management*, 10(2), 175–96.

Kern, F. and Smith, A. 2008. Restructuring Energy Systems for Sustainability? Energy Transition Policy in the Netherlands, *Energy Policy*, 36(11), 4093–103.

Loorbach, D. 2007. *Transition Management: New Mode of Governance for Sustainable Development*. Preprint. Rotterdam: Erasmus University.

Loorbach, D. and Rotmans, J. 2009. Towards a Better Understanding of Transition and their Governance: A Systemic and Reflexive Approach Perspective, in *Transition to Sustainable Development: New Directions in The Study of Long Term Transformative Change*, edited by J. Grin, J. Rotmans and J. Schot. New York and London: Routledge.

Ministry of Economic Affairs. 2004. *Innovation in Energy Policy. Energy Transition: State of Affairs and the Way Ahead.* The Hague: Ministry of Economic Affairs.

Ministry of Housing, Spatial Planning and the Environment (VROM). 2001. *Where there's a Will there's a World. Working on Sustainability. 4th National Environmental Policy Plan – Summary.* The Hague: VROM.

Mol, A. 2000. The Environmental Movement in an Era of Ecological Modernisation, *Geoforum*, 31(1), 45–56.

Nielsen, K.A. and Nielsen, B.S. 2003. *En menneskelig Natur. Aktionsforskning for bæredygtig og politisk kultur.* Copenhagen: Frydenlund.

Nowotny, H., Scott, P. and Gibbons, M. 2003. Mode 2 Revisited: The New Production of Knowledge, *Minerva*, 41(3), 179–94.

Pinch, T.J. and Bijker, W. 1990. The Social Construction of Facts and Artefacts, in *The Social Construction of Technological Systems*, edited by W. Bijker. Cambridge, MA: MIT Press, 17–50.

Pløger, J. 2004. Strife: Urban Planning and Agonism, *Planning Theory*, 3(1), 71–92.

Rhodes, R.A.W. 1997. *Understanding Governance.* Buckingham/Philidelphia: Open University Press.

Rip, A. and Kemp, R. 1998. Technological Change, in *Human Choice and Climate Change, Vol 2: Resources and Technology*, edited by S. Rayner and E.L. Malone. Columbus, OH: Battelle Press, 327–99.

Shove, E. and Walker, G. 2007. CAUTION! Transitions Ahead: Politics, Practice, and Sustainable Transition Management, *Environment and Planning A*, 39(4), 763–70.

Smith, A. 2003. Transforming Technological Regimes for Sustainable Development: A Role for Alternative Technology Niches? *Science and Public Policy*, 30(2), 127–35.

Smith, A. and Kern, F. 2007. The Transition Discourse in the Ecological Modernisation of the Netherlands. *SPRU Electronic Working Paper Series* [Online], 160, 1-24. Available at: http://www.sussex.ac.uk/spru/documents/sewp160.pdf [accessed: 9 November 2009].

Søndergård, B., Hansen, O.E. and Holm, J. 2004. Ecological Modernisation and Institutional Transformations in the Danish Textile Industry, *Journal of Cleaner Production*, 12(4), 337–52.

Vellinga, P. 2001. Industrial Transformation: Towards Sustainability in Production and Consumption Processes. *Newsletter of the International Human Dimensions Programme on Global Environmental Change* [Online], 2001(4). Available at: http://www.uni-bonn.de/ihdp/html/publications/update/IHDPUpdate01_04.html [accessed: 9 November 2009].

Voß, J-P. and Kemp, R. 2006. Sustainability and Reflexive Governance, in *Reflexive Governance for Sustainable Development: Introduction*, edited by J-P. Voß et.al. Cheltenham: Edward Elgar, 3–28.

Walker, W. 2000. Entrapment in Large Technology Systems: Institutional Commitment and Power Relations, *Research Policy*, 29(7/8), 833–46.

Chapter 6

On a Sustainable Chemicals Policy: The Significance of Risk Assessment and REACH

Jette Rank, Kristian Syberg and Lars Carlsen

Introduction

In its most often referred version, the concept of sustainable development is based on considerations regarding the three 'pillars': environmental, social and economical development. Even though the essence of the concept holds all three dimensions, it is sometimes feasible to consider development within one of the three dimensions, since such focus enables a more detailed evaluation of the development of the specific dimension. Since all three 'pillars' are essential for sustainable development, sustainable demands for each of them must be met before sustainable development can be achieved. However, it is important that all three areas are addressed simultaneously both individually and as a whole.

In the present chapter we focus on the environmental dimension of the regulation of chemicals in the new EU regulatory framework, REACH, for the Registration, Evaluation and Authorisation of CHemicals (EC 2006) that was formally adopted by the European Council on 18 December 2006 and which entered into force on 1 June 2007. This approach was chosen because the adverse effects of the most hazardous chemicals threaten one of the fundamental aspects of sustainable development: current and future generations' health and ability to reproduce (Skakkebæk et al. 2006). Furthermore, it is often more costly to address the problems with hazardous chemicals after the contamination has occurred. As an example, the Nordic Council of Ministers (NCM) estimated that the cost of wastewater treatment and remediation of polychlorinated biphenyls (PCBs) in the EU would be in the region of €15 billion due to late action rather than prevention (NCM 2004). The report concludes that if REACH can prevent similar mistakes for only a few chemicals, our society could save billions of euros. This underlines that prevention of exposure to hazardous chemicals would also be an economic benefit to our society. This illustrates the connection between the different 'pillars' and demonstrates that focusing on one dimension influences the other two.

The rapid development of industrial chemicals during the last decades together with constantly increased findings of adverse effects on the environment and human health constitutes the basis of the discussion about the sustainability of the present European chemical policy. The regulation of chemicals in Western countries has obviously been insufficient so far with regard to protecting the environment and human health both on individual and on population levels (EC 2001) and can therefore not be described as sustainable. Unambiguously, the situation in the developing countries appears to be significantly worse in this respect. The most recent demonstration of this is the production of hazardous pesticides in India by Western companies, substances that were forbidden in Europe long ago since they are considered highly toxic to both the environment and to humans (Food and Agriculture Organization (FAO) 2006). Examples of inadequate chemical regulations are numerous and well documented. Rachel Carson was the first to describe the irresponsible use of carcinogenic pesticides (Carson 1962), and followed many years later by Theo Colborn who in the book *Our Stolen Future* focuses on the adverse health effects from endocrine disrupting man-made chemicals (Colborn et al. 1996). Also, the report *Late Lessons from Early Warnings* (EEA 2001) gives examples of well-known harmful chemicals for which there has been no risk reduction in spite of the knowledge of their hazardous potential.

One main obstacle in relation to 'sustainability' appears to be the very high number of chemicals which are in circulation at a global level. Thus, within the European Union, the chemical legislation has, since 1981, allowed the industry to use about 100,000 chemicals included in the European Inventory of Existing Commercial Chemical Substances (EINECS) virtually without restrictions (EINECS 1981). However, even though 'only' 20,000–30,000 of these industrial chemicals are apparently in circulation on the European market, data on environmental fate as well as on the toxic effects of these chemicals both as pure substances and as constituents in various formulations are scarce and in the majority of cases even lacking, or at least not made available by the industry. Thus, because of this data gap, it has been virtually impossible to carry out proper risk assessments for the vast majority of chemicals available on the European market. Furthermore, the existing data are primarily concerned with effects of single substances, and do thus not take any possible synergistic mixture effects with other chemicals into consideration (EC 2003), even though it is generally accepted that chemicals seldom occur alone in the environment (Bachhaus et al. 2003). Consequently, it has been argued that the current use of chemicals should be considered a huge experiment, with nature and humans as the test organisms (Colborn et al.1996).

Because of the comprehensive international reports on the impact of hazardous chemicals, the green movements in the EU were very incensed and in 2000 some key demands for a more responsible EU chemicals policy were made at an international conference in Copenhagen (Boye 2000). The five demands developed by the conference are shown in Table 6.1. Most of them could be part of a sustainable chemicals policy and they have also to a certain degree influenced the process of the REACH proposal. However, to further improve these key demands we would like to add four new demands, which we find mandatory for a sustainable

Table 6.1   Copenhagen chemicals charter: The five key demands for a better EU chemicals policy

1. A full right to know – including what chemicals are present.
2. A deadline by which all chemicals on the market must have had their safety independently assessed. All uses of a chemical should be approved and should be demonstrated to be safe beyond reasonable doubt.
3. A phase out of persistent or bio-accumulative chemicals.
4. A requirement to substitute less safe chemicals with safer alternatives.
5. A commitment to stop all releases to the environment of hazardous substances by 2020.

Source: Boye 2000.

chemicals policy (Table 6.2). The new demands call attention to the importance of all chemicals being included in the regulation and that the legislation must be easy to understand and manage.

It could be argued that because of a very high degree of complexity relating to the risk assessment of industrial chemicals, it appears virtually impossible to reduce the risk to zero. Nevertheless, the overall concept of a sustainable chemicals policy should be defined as 'the use of chemicals without causing any harm to the environmental or to human health throughout the full lifecycle of the chemical, i.e., comprising primary production, down stream manufacturing, use in consumer products, fate in the environment and in the eventual waste phase including disposal as well as, e.g. removal by incineration'. Therefore, a policy trying to reduce the risks arising from dangerous chemicals could be an important contribution to, although not necessarily sufficient for, sustainable development within the complex use of chemicals in production, consumption and waste handling.

In the present chapter the regulatory framework provided by REACH will be discussed in the light of sustainability, as one specific objective for the new chemical regulation is to achieve sustainable development by providing a high level of protection of human health and the environment (EC 2006). Finally, the analysis will focus on the question of whether REACH will serve its purpose with regard to the demands listed in Table 6.2 or not.

Table 6.2   Four demands to a sustainable chemicals policy

1. A chemical legislation that includes all chemicals.
2. A legislation simple and easy for all users to manage.
3. A valid risk assessment of all chemicals should be provided on the basis of a full set of toxicological data and exposure data.
4. An appropriate risk reducing measure aiming at a risk level as close to zero as possible.

REACH

The overall objective of the new EU regulatory framework for the Registration, Evaluation and Authorisation of CHemicals (REACH) is to ensure an effective functioning of the common market for chemical substances, and at the same time protect the environment and human health against adverse effects caused by the manufacture and use of chemicals under reasonably foreseeable conditions. The REACH framework consists of four elements: (1) registration; (2) evaluation; and (3) authorisation of chemicals; in addition to (4) possible restrictions (e.g. substitution of hazardous chemicals with less hazardous or complete ban) of specific chemicals.

According to REACH, it is mandatory to submit a registration for all chemicals produced or imported[1] in amounts larger than 1 tpy (ton per year) per producer (EC 2006). The necessary data must be provided by the manufacturers and importers (see below concerning the data requirements) in addition to recommendations regarding appropriate risk management measures.

The information requirements are tiered by tonnage. Thus, for chemicals produced or imported in 1 to 10 tpy only a few data are required. The most important are the physical and chemical data and a few toxicology data, and a full data package should only be submitted if there is an indication that the substance should be classified as dangerous and will be used in consumer products or commercially (EC 2006). A chemical safety report has to be provided for substances produced or imported in more than 10 tpy. This report must include:

- human health hazard assessment;
- human health hazard assessment of physiochemical properties;
- environmental hazard assessment;
- assessment of persistent, bioaccumulating and toxic (PBT), and very persistent and very bioaccumulating (vPvB) characteristics;
- assessment for possible restrictions for certain uses (EC 2003).

If a substance meets the criteria for classification as dangerous in accordance with Directive 67/548/EEC (EEC 1967) or is assessed to be PBT and vPvB, the assessment must include an exposure assessment as well as a risk characterisation. For chemicals produced in more than 100 tpy per producer an evaluation is mandatory. For chemical substances of very high concern, i.e., the so-called CMR substances (carcinogens, mutagens and substances toxic to reproduction), endocrine disrupting chemicals (EDCs), persistent organic pollutants (POPs) and substances that are persistent, bioaccumulating and toxic (PBTs) or very persistent and very bioaccumulating (vPvBs), authorisation must be obtained for selected substances in order to use them (EC 2006: Article 57). Priority will be given to those chemicals being PBT, vPvB or having wide dispersive use or are produced in high volumes

---

1   Imported substances are those that are produced outside the borders of the European Community.

(EC 2006: Article 58). The authorisation system has been considered one of the most important changes in the new regulation since it is supposed to encourage the producers to substitute the most hazardous chemicals (EC 2001, Warhust 2006). However, it still remains to be disclosed how the practical use will be managed.

In contrast to the former chemical regulation where risk assessments were preformed by the member states, the industry will be responsible for the risk assessment. This new responsibility is considered a very important part of REACH (Führ and Bizer 2007). The industry is forced to deliver detailed chemical safety reports based on a life cycle perspective and it is compelled to propose a risk characterisation and classification for each chemical produced in a volume more than 10 tpy. The subsequent substance evaluations based on these risk assessments will be drawn up by the member states. However, it must be remembered that according to the authorisation system of REACH (EC 2006: Article 57 and 59), CMR substances and endocrine disrupting chemicals, for which it is not possible to determine a lower threshold for their action, or for substances having PBT or vPvB properties, authorisation may not be granted unless it could be argued that socio-economic factors relating to the chemicals are of great importance. Clearly, the PBT and vPvB substances substitution should be required because these substances will remain uncontrolled in the environment for a very long time with the possibility of accumulating in the food chains and concentrating in higher animals such as polar bears and humans. Thus, the absolute refusals of granting authorisation in the PBT and vPvB substances would unequivocally be in line with thinking in terms of sustainability. However, apparently this is not the case within the REACH framework.

The technical, scientific and administrative aspects of REACH will be handled by the new European Chemical Agency (ECHA) in Helsinki, which will also be responsible for the development of the evaluation criteria. It will also be responsible for making the list of chemicals for authorisation. The first was published in June 2008 (ECHA 2008) and it is expected that socio-economic arguments will be an important part of such an application.

Obviously REACH does not include all chemicals and does as such not a priori comply with the above requirements for a sustainable chemicals policy. Only chemicals produced or imported in more than 1 tpy are considered to be of concern, which apparently eliminates approximately 70 per cent of the more than 100,000 substances registered on the EINECS list. This limit is arbitrary and might as well have been lower. The explanation for this should be found in economic arguments rather than arguments of a more scientific nature.

In addition, the significantly reduced requirement for registration, i.e. the exempting from full health and safety test of substances produced or imported in amounts between 1 and 10 tpy will further reduce the number of substances to be risk assessed by approximately 17,000. However, some of these chemicals produced in smaller amounts could very well turn up to be highly toxic or very persistent as hazardous characteristics obviously are not linked to the production volume. In this case, the costs have also outweighed the more scientific approach of risk assessment and, thus, in this respect, REACH does not qualify as a sustainable chemicals policy.

Table 6.3   **Assessment factors and matching data for the aquatic environment according to the Technical Guidance Document (TGD 2003)**

Factor 1000:	At least one short-term LC_{50} or EC_{50} from each of three trophic levels of the base set (fish, daphnia and algae).
Factor 100:	One long-term NOEC (either fish or Daphnia).
Factor 50:	Two long term NOECs from species representing two trophic levels (fish and/or daphnia and/or algae).
Factor 10:	Long-term NOECs from at least three species (normally fish, Daphnia and algae) representing three trophic levels.
Others:	If field data or model ecosystem data are available, application factors are reviewed on a case by case basis.

Note: LC50 means Lethal Concentration for 50 per cent of the test organisms, EC50 means Effect Concentration for 50 per cent of the test organisms, NOEC means No Observed Effect Concentration.

Risk Assessment and Sustainability

The environmental risk assessment of a chemical consists of four steps: (1) hazard identification; (2) dose response assessment; (3) exposure assessment; and (4) risk characterisation (EC 1994). The aim of the hazard identification is to identify the effects of concern and to propose a classification of the substance, and perform a dose response assessment. For the purpose of environmental risk assessment, the no observed effect concentration (NOEC) will be determined and from that the predicted no effect concentration (PNEC) can be derived by the use of assessment factors (see Table 6.3). In human risk assessment it is not the exposure concentration which is of interest but the internal dose given to the test animals. Therefore, a no observed effect level (NOEL) will be used in the risk assessment, often altered to a derived no observed level (DNEL) (EC 2003).

In principle, the environmental exposure assessment should consider the following stages of the life cycle of the substance: production, processing, transportation and storage, formulation, (comprising blending and mixing of substances in preparations), use (comprising large-scale use (industry) and/or professional small-scale use (trade) and/or private or consumer use), and disposal and waste treatment, e.g. incineration and recycling. However, in practice, generic exposure scenarios might be used for a range of the chemicals in the actual implementation of REACH (ECB 2007). The objective of the exposure assessment is to derive the concentration of the chemical in the environment, the so-called predicted environmental concentration (PEC) for the chemical. It is not clear if the actual implementation of the demands to the exposure scenarios will eventually strengthen or weaken the predictive power of the risk assessments.

In the final step, the risk characterisation, the so-called risk coefficient, RQ = PEC/PNEC, is derived. A priori, it is believed that for RQ < 1 the chemical does not possess a risk, whereas for RQ ≥ 1 the substance possesses a risk. However, it is generally accepted that it is not possible to handle possible risk that rigidly, and other data such as persistence and bioaccumulation will also be a part of the assessment.

In order to account for any uncertainties associated with the risk quotient for the single substances, so-called assessment factors are applied when NOEC is changed to PNEC (see Table 6.3). The principle for using assessment factors is based on the concept that few data will result in the use of a high factor and therefore expresses a high degree of uncertainty. In the European Technical Guidance Document (TGD), assessment factors in the range from 1,000 to 10 must be applied for organisms living in fresh water, soil and sediment. However, for the marine environment, which is regarded to be very sensitive, a factor of 10,000 can be used if only few data are available (EC 2003).

Obviously, the rather simple approach using the risk quotient higher or lower than one as a delimiter for potential hazard constitutes, a priori, a problem. Thus, the immediate question is if a compound exhibiting a RQ = 1.05 is really hazardous, whereas a compound with RQ = 0.95 is not. When applying the lower assessment factors this becomes increasingly crucial. This will often be the case as the requirement for using an assessment factor equal to 10 or lower will prevail. Bearing this in mind, it seems as though the present risk assessment process cannot automatically be regarded as sustainable. This would call for a more subtle view. One, still simple, possibility in this context could be to introduce a range, e.g. 0.1 < RQ <10, for which further testing would be mandatory to verify or exclude potential hazards. This would automatically introduce a further safety margin of a factor 10. If it is not possible to verify that a substance does not possess a hazard within this range then the precautionary principle could be applied. This could lead to restrictions or even ban on the use of the substance until further evidence could be presented.

Here we have primarily described the risk assessment for the environment, but it is, in principle, similar to the risk assessment for human health as the exposure level (EL) is compared to the derived no effect level (DNEL) for the most sensitive organism, the most sensitive endpoint and the most sensitive effect study. A margin of safety (MOS) is calculated: MOS = DNEL/EL without using any assessment factors. However, the MOS value could, in principle, be compared with an assessment factor because the consensus is that the quotient should be more than 100 in order to be safe, and obviously it should be as high as possible.

For both the environmental and the human health risk assessments the exposure is normally regarded the most uncertain factor. However, it may also be questioned if all kinds of effects have been considered. Thus, e.g. the case of endocrine disrupting chemicals has shown that we still need new methods of testing to be sure that the risk assessment procedure provides all relevant toxicology data. This is important because we suspect some hormone mimicking chemicals to cause harm on both environment and human health (Colborn et al. 1996), and increasing human diseases such as testicular cancer, hypospadias, cryptorchism and reduced sperm quality are suspected to be linked to the action of certain chemicals (Rank 2005, Skakkebæk et al. 2006). Therefore, availability of guidelines to relevant test methods in this field is important if the evaluation and authorisation of these substances should be carried out in a safe way. For the forthcoming period awaiting the development of appropriate test procedures, the precautionary principle should be brought into play as it is described in the communication paper from the Commission (EC 2000) and further recommended in REACH (EC 2006).

The main problem with the standardised risk assessment is that it is based on exposure evaluation because, as it is argued, without exposure or exposure below a certain limit no harm will happen. However, considering that most industrial chemicals have a rather uncontrolled environmental fate leading to an insecure exposure evaluation, it is obvious that no risk assessment will lead to a sustainable use of a chemical as it is virtually impossible to derive scenarios with a calculated zero exposure. An example of this is the risk from the phthalate DEHP (Rank 2005). This chemical can be measured in all environmental compartments, all kinds of food (Wormuth et al. 2006) and in human blood because it seeps out from the products and its spread is uncontrolled. Such chemicals can never be enclosed in a 'zero exposure scenario' as the exposure evaluation is too complex.

Regulation of Chemicals of High Concern

One of the major principles of REACH is that substances of very high concern (SVHC) have to be authorised for specific use (EC 2006: Title VII). It is estimated that approximately 1,500 chemicals fall within this group. It is worthwhile noting that the principle does not imply that the use of the most hazardous chemicals will be banned. In a historical perspective this is worrying since earlier experience with chemicals like the PCBs have unambiguously demonstrated that it is very difficult to ensure that chemicals do not leak into the environment. Thus, history has shown that chemicals with PBT properties tend to accumulate in the environment.

One important aspect which leads REACH in a sustainable direction is the fact that chemicals of very high concern have to be substituted with less dangerous chemicals if these are available at an economic cost (EC 2006: Article 60). This principle is very much in accordance with the essence of sustainability. However, some loopholes are immediately noted. Thus, the actual implementation of the principle suffers from the fact that it is the companies themselves that shall produce substitution plans. This kind of self-regulation has a long history of producing inadequate regulation and it is therefore questionable to what degree the principle can ensure a sustainable chemical regulation. Secondly, the term 'at an economic cost' may very well lead to discussions between the industry and the authorities that eventually may result in an outcome which is not sustainable.

It is stated in REACH that authorisation can be granted if socio-economic benefits outweigh the risk and no suitable alternative exists (EC 2006: Article 60).

The authorisation procedure was expected to be the cornerstone in the new regulation, but could turn out to be a step in the wrong direction in regard to ensuring a sustainable use of chemicals as it may be expected that the socio-economic factors in many cases will be given priority over data pointing at environmental and human health risks (see below).

A further questionable aspect is the determination of the chemical's hazardous properties. In order to make the task of producing data easier for the industry, the amount of required data vary depending on production volume as already described. From a scientific point of view, it is highly questionable that production volume is the only parameter determining what kinds of data are needed. The limit for inclusion in REACH is that the production volume is more than 1 tpy per producer or importer, and very few toxicological data are required for chemicals between 1 and 10 tpy. This may well constitute a significant problem as the hazard of a chemical obviously does not only depend on the volume used but also on the toxic properties of the chemical.

One of the major improvements expected through the implementation of REACH is that the chemical regulation will not solely be based on risk assessments but take into account some inherent chemical properties such as bioaccumulation and persistence. This is a major advantage from an environmental and human health perspective since it recognises that even when exposure scenarios indicate that the chemical poses no actual risk, we cannot be sure that this is actually the case. This uncertainty, especially in relation to the most hazardous chemicals, is unacceptable in a system that imposes sustainable chemical regulation, and it is thus creditable that chemicals regulated under REACH in principle shall be evaluated in regard to such inherent hazardous properties. In order to evaluate whether the criteria in practice impose sustainable chemicals regulation, it is, however, vital to evaluate the specific definition of the criteria.

PBT and vPvB

History has disclosed that chemicals that stay in the environment for a long time and tend to be concentrated in animals pose a major risk. Due to this historical lesson the PBT and vPvB criteria have been constructed. A PBT chemical is one that is Persistent, Bioaccumulating *and* Toxic (see Table 6.4). In essence, chemicals possessing such properties have to be removed from the market. In that sense it must be acknowledged as a step towards sustainability that these criteria have been integrated in REACH. However, whether the actual definition of the three properties can be said to protect the environment and humans from chemicals with the unwanted properties is, on the other hand, highly questionable. Chemicals with a half-life > 60 d in seawater or > 40 d in freshwater or > 180 d in marine sediment or > 120 d freshwater sediment or >120 d in soil are considered persistent (EC 2003). However, for chemicals that are continuously leached to the environment and also are relatively persistent (DDT, PCBs, PAHs, PFOS, etc.), this will mean that concentrations in environmental compartments as soils or sediments will increase over time.

Table 6.4   Criteria for the identification of persistent, bioaccumulative and toxic substances

---

Persistency: t½ > 60 days in marine water or
                  > 40 days in fresh or estuarine water or
                  >180 days in marine sediment or
                  > 120 days in fresh water sediment
                  > 120 days in soil

Bio-Concentration Factor (BCF): > 2000

Toxicity: Long-term NOEC < 0.01 mg/l for marine or fresh water or
              Classification as Carcinogenic, Mutagenic or toxic for Reproduction or
              Evidence of chronic toxicity

It is well documented that hazardous chemicals in the sediment pose a risk not only to the organisms that live in the sediment, but also to the entire aquatic food chain (Colborn et al. 1996, Maruya and Lee 1998, Palmqvist et al. 2006). In order to be characterised as bioaccumulating in a risk assessment context, a chemical has to be concentrated in the organism with more than a factor of 2,000 compared with the surrounding environment (see Table 6.4). Taking into consideration that we know that bioaccumulating chemicals pose a major risk, especially to animals in the top of the food chain including humans, it is difficult to see how chemicals that bioaccumulate a thousand or even a hundred times may not pose a potential risk to organisms as well as to future generations. Moreover, chemicals have to be toxic in order to be classified as PBT (see Table 6.4).

Further, a historical retrospect discloses that we have been surprised several times by chemicals with new unexpected toxic properties that were not originally considered (Sanderson et al. 2003, Rahman et al. 2001). To assume that we, by implementing the PBT criteria in REACH, will focus on all known and unknown toxicities associated with the substance is far beyond the reality. This is the reason why the vPvB criterion was constructed. In regard to this criterion, the chemical does not need to have known toxic properties, but 'only' needs to be very persistent and very bioaccumulating. However, compared to the PBT criteria shown in Table 6.4, chemicals have to be more persistent and more bioaccumulating in order to be classified as vPvB. Thus, compounds with half lives > 60 d in seawater or freshwater or > 180 d in marine or freshwater sediment or >180 d in soil are considered vP, and the bioconcentration factor > 5000 will lead to a classification as vB.

In respect to evaluating whether the PBT or vPvB criteria ensure a sustainable management of chemicals, it could be asked if the principles are in accordance with our definition of sustainable chemical regulation. Obviously, the criteria are constructed in such a way that it does not necessarily protect us from all persistent and bioaccumulating chemicals and does thus not ensure a sustainable chemical regulation. Moreover, it should be considered that the here mentioned criteria for persistence and bioaccumulation are not scientific but more practical limits, which could be changed if the regulatory authorities want to strengthen or weaken the regulation.

As mentioned previously, authorisation should not be granted for substances having PBT or vPvB properties (EC 2006: Article 57 and 59). Thus, here the REACH concept is in line with sustainability.

Chemicals with Specific Hazardous Properties (CMR and ED)

A special group of chemicals with unwanted inherent hazardous properties are the CMR chemicals, i.e. chemicals, which are Carcinogenic, Mutagenic or toxic for Reproduction (EC 2003). Carcinogenic and mutagenic chemicals may have serious negative impacts on human health including cancer and have thus been prioritised for many years. Chemicals that are toxic to reproduction are very important in relation to hazards to future generations and thus with regard to sustainability. Unfortunately, regulation of CMR chemicals under REACH is somewhat inconsequent because authorisation can be given if the producer can show that the risk can be 'adequately controlled' and that a safe threshold with regard to human body load is not surpassed. This means that these potentially highly hazardous inherent properties will not be sufficient to ban a chemical, but that exposure scenarios have to be considered.

With the very limited knowledge we have about the actual use of many chemicals such exposure scenarios are at best very difficult to conduct properly and at worst inadequate. Further, the term 'adequately controlled' (EC 2006: Article 60) constitutes a major loophole in the REACH system, since no definition as to 'adequate' is offered. Furthermore, it is highly questionable whether safe thresholds can be determined, especially taking into consideration that mutagenic or carcinogenic chemicals, which act directly on the DNA, are not expected to have limit values. Moreover, the chemicals are assessed individually without regard to their possible combined effect with other chemicals. It is well documented that such mixture may exhibit significant toxic effects (Greco et al. 1995, Altenburger et al. 2000, Silva et al. 2002, Backhaus et al. 2003), and several investigations have verified that we have a large number of chemicals in our bodies (Colburn et al.1996, CDC 2005). Hence, it is questionable if a chemical legislation, which does not fully protect us from CMR chemicals, can ensure sustainability. However, it should in context be recalled that REACH does state that authorisation should not be granted for substances being carcinogenic, mutagenic and toxic or exhibiting endocrine disruption properties for which it is not possible to determine a lower threshold for their action (EC 2006: Article 57 and 59). Nevertheless, a proof for such a lacking threshold may be hard to obtain. Thus, this calls again for bringing the precautionary principle into play. A similar argument can be made in regard to the endocrine disrupting chemicals (EDCs). This latest example of new hazardous effects poses a significant risk to both environment and human health. Those chemicals that mimic the action of sex hormones receive special attention. It is thus good news that the chemicals are considered as of very high concern, but the actual regulation does only apply when there is scientific evidence of probable serious effects to human health and the environment (EC 2006). Our knowledge of these chemicals is in its infancy, so it is questionable to assume that we can determine when these chemicals do not lead to serious effects. Furthermore, the assessments are again based on single chemical effects without taking possible mixture effects into consideration. It has been demonstrated that EDCs can have mixture effects that are significantly higher that those observed for the single chemicals (Silva et al. 2002). If REACH should be sustainable in regard to regulation of these chemicals, they should therefore not be allowed at all. In conclusion it appears that the regulation of these high-risk chemicals cannot be described as sustainable.

Responsibility and Socio-economic Factors

Without any doubt REACH represents a new regulatory approach. The producers, importers and downstream users will be responsible for the classification of the chemicals and the risk assessment, which was previously placed on the regulatory bodies. The new concept could be called a paradigm shift and has important implications (Führ and Bizer 2007). Taking into consideration that the producers and downstream users on the basis of their own risk assessment should also develop measures of risk reduction, some guidelines on this kind of behaviour should be created. Moreover, the authorisation system is also a new approach where some interesting challenges exist, mostly because REACH, as described earlier, is born with inherent loopholes for the industry.

A further major obstacle possibly undermining a sustainable use of chemicals may well be socio-economic factors. According to REACH, the authorisation of chemicals covered by Articles 55–59 in REACH (EC 2006) can be granted if the manufacturer, through a socio-economic analysis, can show that, e.g., the socio-economic benefits outweigh the risks (EC 2006: Article 60). It is worthwhile noting that REACH does not include recommendations about how to balance between the risks on the one hand and the socio-economic analysis on the other hand.

If a chemical is classified in a category which needs authorisation, the permission is not solely given in accordance with the hazardous properties of the chemical. A socio-economic analysis (SEA) should be carried out and the results be used in the final decision. If the societal benefits of using the chemical outweigh the impacts on the environment and human health, it will likely result in an authorisation. How such an SEA will be carried out is not very clear, but according to REACH, the SEA among other things could include assessment of (EC 2006: Annex XVI):

• impact on the applicant, industry or all other actors in the supply chain;
• impacts on consumers, for example product prices, changes in composition and consumer choice;

- social implications, for example job security and employment;
- economic consequences of availability suitability, and technical feasibility of alternatives;
- wider implications on trade, competition and economic development;
- costs linked to alternative risk management measures;
- social and economic benefits of the proposed restriction on, for example, worker health, the environment and the distribution of these benefits.

It must in this connection be emphasised that guidelines for what this socio-economic analysis could comprise are currently under development (EC 2006: Annex XVI) and are not available yet (May 2010).However, it is a problem that the applicant of the authorisation will make the SEA as the objectivity of the analysis could be questioned. It is clear that a company applying for an authorisation for a certain chemical will try to emphasise that the economic consequences on the involved industry, downstream users and consumers of not using the substance are much higher than the impact on the environment and human health otherwise they would not try to obtain an authorisation. It is also worthwhile noting that REACH does not include recommendations about how to balance between the risks on the one hand and the SEA on the other hand. But, hopefully, this may be included in the coming guidelines. As described earlier, it seems as though the socio-economic analysis could undermine the whole authorisation system. To handle these problems a special SEA committee shall be established and when a SEA is part of the application, the committee will evaluate the SEA and form an opinion taking all aspects into account.

Towards a Global Sustainable Chemical Policy

As chemicals do not recognise borders they constitute a global problem, and the concern regarding hazardous chemicals has, on all levels, significantly increased over the last two decades. To obtain a more sustainable use of chemicals, international meetings have had an impact on the use of the most dangerous chemicals placing them high on the agenda, and many important conventions have been made and ratified (Table 6.5).

At the world summit in Johannesburg in 2002, it was recommended that before 2020 a sound management of chemicals through their whole life cycle should be implemented to reduce the effect on humans and the environment. Further, a new classification system, the Globally Harmonized System of Classification and Labelling of Chemicals (GHS) was launched by the United Nations Economic Commission for Europe (UNECE) (UNECE 2007). The idea was that GHS should be implemented at the same time as REACH and be in force not later than 2008. GHS is meant to create a high degree of safety when hazardous chemicals are handled at all levels all over the world. GHS allows a uniform development of

Table 6.5   International conventions of importance to the pollution prevention, chemical policy and management

Name	Year of creation	Purpose of the convention
Aarhus Convention	1998	Convention on access to information, public participation in decision-making and access to justice in environmental matters.
Montreal Convention	1987	Montreal protocol on substances that deplete the ozone layer.
Rotterdam Convention	2004	The convention should promote shared responsibility in the international trade of certain hazardous chemicals and pesticides in order to protect human health and environment.
Basel Convention	2005	Convention on the control of transboundary movements of hazardous wastes and their disposal.
Stockholm Convention	1997	To protect human health and the environment from persistent organic pollutants (POPs).
SAICM	2006	Strategic approach to international chemicals management.

national policies, but still remains flexible enough to accommodate any special requirements that have to be met.

The Rotterdam Convention on the Prior Informed Consent Procedure for Certain Hazardous Chemicals and Pesticides in International Trade, the so-called PIC Convention (PIC 2004), focuses on shared responsibility when banned or severely restricted pesticides and industrial chemicals are part of international trade. The purpose is to protect human health and the environment from potential harm by the exchange of information on the hazardous properties. To ensure this intention is fulfilled, the Convention makes a list of chemicals which are subject to 'the prior informed consent procedure' where the risk using the specific chemical is described. However, the PIC list of chemicals is not very comprehensive and most compounds are pesticides, which are already banned.

In a global context, the Stockholm Convention on Persistent Organic Pollutants (POPs) (UNEP 2001), often called the POP Convention, is very important. The purpose of this is to make an international ban to protect humans and the environment against the most hazardous POPs. Until now only 12 chemicals, the so-called dirty dozen, or groups of chemicals, have been put on the list. Some of these are also part of the Rotterdam Convention, e.g. the pesticides aldrin, dieldrin and DDT and the industrial chemical PCB. Being on the POP Convention list means that the chemicals are banned globally. However, although covered by the POP Convention, the use of DDT is not completely banned as the substance can be used for disease vector control in selected parts of the world where malaria constitutes a threat against human health. Even taking this fact into account, the use of DDT as a pesticide can therefore not be part of sustainable development, as the poison is highly toxic, e.g. towards bird reproduction as it causes decreasing thickness of the egg shells. This was already described in The Silent Spring (Carson 1961), where the title referred to the fact that birds stopped singing in areas where DDT was used extensively against insects.

It is obvious that a significant effort on a global level is being made and most probably other persistent organic pollutants like brominated flame retardants, as e.g. the so called PBDEs, and the very persistent fluorinated chemicals like PFOS, and possibly selected highly persistent polycyclic aromatic compounds, are obvious candidates. Therefore, to contribute to sustainable development, the POP Convention needs to be updated with a 'second generation' of hazardous persistent organic chemicals.

In general, consensus about the safe and more sustainable usage of chemicals must take place on a global level prevails. The newest convention is the Strategic Approach to International Chemicals Management (SAICM) (UNEP 2006), which aims to support risk reduction from chemicals via knowledge and information, governance, capacity-building and technical cooperation, and focus on illegal international traffic (e.g. sales of banned CFCs from Asian countries in packing having false labels). It is a global plan of action and it is interesting to follow this new initiative and see if it will contribute to a safer and more sustainable management of chemicals.

During the process of this international work, it seems as though some kind of consensus could be obtained about how to evaluate the hazards and risks of chemicals. The European risk assessment system is in that respect not different from the risk assessment system in other Western countries. The new GHS is also a step towards international consensus about risk evaluation of chemicals. It could be expected that REACH by providing new toxic data in the future will push global efforts in the right direction by filling in the information gap concerning existing chemicals (Warhust 2006).

Unfortunately, even if an international consensus is obtained about certain chemicals, there is still no guarantee that the hazardous chemicals will be banned or used in a sustainable way, because all the international conventions can be seen to some extent as declarations of intent. Thus, there will be little or no consequences if the ratified agreements are not followed. However, it should in this context be mentioned that the Stockholm Convention, i.e., the so-called POP Convention including 'the dirty dozen' can unequivocally be regarded as a step in the right direction, i.e. showing the chemical industry that some chemicals of high concern can be globally banned and we can only hope that the industry will bear that in mind when developing new substances.

REACH and the Proposed Demands

The chemicals problem has to be regarded as being of transboundary character and should therefore be dealt with globally. However, many important initiatives to stop the unsafe use of chemicals are deployed at lower levels and REACH, belonging to this category, is regarded the most radical regulation of chemicals in the world, and it is the hope, even though it is not 'perfect', that it will have a positive influence on chemical policy and chemical management at a global level.

However, as described in the present chapter, REACH is not a guarantee for a complete risk-free use of chemicals. Many loopholes have to be filled before it can be claimed that this new regulation contributes to a proper sustainable development. Already, the authorisation system seems to be less restrictive than originally expected, as only 16 substances were found to be considered as substances of very high concern on the first list of chemicals proposed for authorisation (ECHA 2008). Furthermore, before any authorisation is given or refused there is a consultation period, in which it is possible for the industry and other interested parties to lobby in order to play down and minimise the hazardous properties of the substances. When the chemicals are eventually placed on the list, the producer or importer should apply for authorisation, and thus it could well be expected that the socio-economic arguments will be the most significant part of such an application.

According to the first of the new demands put forward in this chapter (Table 6.2), all chemicals should be part of the regulation. However, the many chemicals produced in less than 1 tpy per producer are not included in REACH, and as toxicity is not related to tonnage these chemicals could of course be very hazardous. In the near future, we will recommend gaining knowledge of the hazardous properties of those chemicals by carrying out some simple computer-aided assessments with QSAR or alike. This could be done at a low cost.

The second demand about a simple and understandable legislation is not fulfilled. Even though the legal text is clearly written, there are many unclear parts that need interpretation. In particular, the authorisation system is unclear. The term 'adequate control' is so far not explained, and the socio-economic analysis will not be described in detail because it is up to the applicants of authorisation to decide what they find important to investigate.

The third demand of a valid risk assessment of all chemicals is not at all part of REACH, as at least 20,000 chemicals (< 1 tpy) and 17,000 chemicals produced between 1 and 10 tpy will not go through a risk assessment. Therefore, it is not possible to fulfil the fourth point demanding risk reducing measures aimed at the lowest risk levels obtainable. In conclusion REACH, even though it is a step forward, is far from being a sustainable chemicals policy.

References

Altenburger, R., Backhaus, T., Boedeker, W., Faust, M., Scholze, M. and Grimme, L.H. 2000. Predictability of the Toxicity of Multiple Chemical Mixtures to Vibrio Fischeri: Mixtures Composed of Similarly Acting Chemicals. *Environmental Toxicology and Chemistry*, 19(9), 2341–7.
Carson, R. 1962. *Silent Spring*. London: Penguin.

CDC (Centers for Disease Control and Prevention). 2005. *Third National Report on Human Exposure to Environmental Chemicals*. Atlanta, GA: National Center for Environmental Health, Division of Laboratory Science.

Colborn, T., Dumanoski, D. and Myers, J. 1996. *Our Stolen Future: Are we Threatening Fertility, Intelligence & Survival? A Scientific Detective Story*. New York: Penguin.

Backhaus, T., Altenburger, R., Arrhenius, Å., Blanck, H., Faust, M., Finizio, A., Gramatica, P., Grote, M., Junghans, M., Meyer, W., Pavan, M., Porsbring, T., Scholze, M., Todeschini, R., Vighi, M., Walter, H. and Grimme, L.H. 2003. The BEAM-project: Prediction and Assessment of Mixture Toxicities in the Aquatic Environment. *Continental Shelf Research*, 23(17–19), 1757–69.

Boye, M. 2000. *Copenhagen Chemicals Charter*. Copenhagen: The European Bureau, The European Consumers' Organization, The Danish Consumer Council, The Danish Society for the Conservation of Nature, and The Danish Ecological Council.

EC. 1994. *Commission Regulation (EC) No 1488/94 of 28 June 1998 Laying Down the Principles for the Assessment of Risk to Man and the Environment of Existing Substances in Accordance with Council Regulation* (EEC) No793/93.

EC. 2000. *Communication from the Commission on the Precautionary Principle*. Brussels: European Commission.

EC. 2001. *The White Paper on Strategy for a Future Chemicals Policy*. COM (2001) 88 Final. Brussels: Commission of the European Communities.

EC. 2003. *2nd Edition of the Technical Guidance Document in Support of Commission Directive 93/67/EEC on Risk Assessment for New Notified Substances*. Commission Regulation (EC) No 1488/94 on Risk Assessment for Existing Substances and Directive 98/8/EC of the European Parliament and of the Council Concerning the Placing of Biocidal Products on the Market. Luxembourg: Office for Official Publications of the European Communities.

EC. 2006. *Regulation (EC) No 1907/2006 of the European Parliament and of the Council of 18 December 2006 Concerning the Registration, Evaluation, Authorisation and Restriction of Chemicals (REACH), establishing a European Chemicals Agency, amending Directive 1999/45/EC and repealing Council Regulation (EEC) No 793/93 and Commission Regulation (EC) No 1488/94 as well as Council Directive 76/769/EEC and Commission Directives 91/155/ EEC, 93/67/EEC, 93/105/EC and 2000/21/EC.* Official Journal of the European Union [Online]. Available at: http://eur-lex.europa.eu/LexUriServ/LexUriServ. do?uri=OJ:L:2006:396:0001:0849:EN:PDF [accessed: 9 November 2009].

ECB (European Chemical Bureau). 2007. *REACH Implementation Project 3.2.2 (RIP 3.2.2). Guidance on Preparing the Chemical Safety Report (CSR), Preliminary Activities Relating to Exposure Scenarios and Exposure Assessment*. [Online]. Available at: http://ecb.jrc.it/reach/rip/ [accessed: August 2007].

ECHA (European Chemicals Agency). 2008. *Candidate List of Substances of Very High Concern for Authorisation*. [Online]. Available at: http://echa.europa.

eu/chem_data/authorisation_process/candidate_list_table_en.asp   [accessed: August 2008].

EEA (European Environment Agency). 2001. *Late Lessons from Early Warnings: The Precautionary Principle 1986–2000*, edited by P. Harremoës, D. Gee, M. MacGarvin, A. Stirling, J. Keys, B. Wynne, S.G. Vaz. Copenhagen: European Environment Agency.

EEC. 1967. *Council Directive 67/548/EEC of 27 June 1967 on the Approximation of Laws, Regulations and Administrative Provisions Relating to the Classification, Packaging and Labelling of Dangerous Substances, as well as later adopted annexes*. Official Journal of the European Union P 196. [Online]. Available at: http://ecb.jrc.it/legislation/1967L0548EC.pdf [accessed: 9 November 2009].

EINECS. 1981. *European Inventory of Existing Commercial Chemical Substances*. [Online]. Available at: http://ecb.jrc.ec.europa.eu/esis/index.php?PGM=ein [accessed: February 2007].

FAO. 2006. *FAO Encourages Early Withdrawal of Highly Toxic Pesticides*. [Online]. Available at: http://www.fao.org/newsroom/en/news/2006/1000471/index.html [accessed: February 2007].

Führ, M. and Bizer, K. 2007. REACH as a Paradigm Shift in Chemical Policy – Responsive Regulation and Behavioral Models. *Journal of Cleaner Production*, 15(4), 327–34.

Greco, W.R., Bravo, G. and Parsons, J.C. 1995. The Search for Synergy: A Critical Review from a Response Surface Perspective. *Pharmacological Reviews*, 47, 332–82.

Maruya, K.A. and Lee, R.E. 1998. Biota-sediment Accumulation and Trophic Transfer Factors for Extremely Hydrophobic Polychlorinated Biphenyls. *Environmental Toxicology and Chemistry*, 17(12), 2463–9.

NCM. 2004. *Cost of Late Action – The Case of PCB*. TemaNord 2004:556. Copenhagen: Nordic Council of Ministers (NCM).

Palmqvist, A., Rasmussen, L.J. and Forbes, V.E. 2006. Influence of Biotransformation on Trophic Transfer of the PAK Fluoranthene. *Aquatic Toxicology*, 80(3), 309–19.

PIC. 2004. *The Rotterdam Convention on the Prior Informed Consent Procedure for Certain Hazardous Chemicals and Pesticides in International Trade*. Geneva: Rotterdam Convention Secretariat, United Nations Environment Programme (UNEP).

Rahman, F., Langford, K.H., Scrimshaw, M.D. and Lester, J.N. 2001. Polybrominated Diphenyl Ether (PBDE) Flame Retardants. *Science of the Total Environment*, 275(1–3), 1–17.

Sanderson, H., Boudreau, T.M., Mabury, S.A. and Solomon, K.R. 2003. Impact of Perfluorooctonic Acid on the Structure of the Zooplankton Community in Indoor Microcosms. *Aquatic Toxicology*, 62(3), 227–34.

Silva, E., Rajapakse, N. and Kortenkamp, A. 2002. Something from 'Nothing' – Eight Weak Chemicals Combined at Concentrations Below NOEC's Produce

Significant Mixture Effects. *Environmental Science and Technology*, 36(8), 1751–8.

Skakkebæk, N.E., Jørgensen, N., Main, K.M., Meyts, E.R.-D., Leffers, H., Andersson, A.-M., Juul, A., Carlsen, E., Mortensen, G.K., Jensen, T.K. and Toppari, J. 2006. Is Human Fecundity Declining? *International Journal of Andrology*, 29(1), 2–11.

UNECE (The United Nations Economic Commission for Europe). 2007. *Globally Harmonized System of Classification and Labelling of Chemicals*. Second Revised Edition. New York and Geneva: United Nations.

UNEP (United Nations Environment Programme). 2001. *Stockholm Convention on Persistent Organic Pollutants*. [Online]. Available at: http://chm.pops.int/ [accessed: February 2007].

UNEP (United Nations Environment Programme). 2006. *Strategic Approach to International Chemicals Management*. [Online]. Available at: http://www.saicm.org [accessed: February 2007].

Warhust, A.M. 2006. Assessing and Managing the Hazards and Risks of Chemicals in the Real World – The Role of EU's REACH Proposal in Future Regulation of Chemicals. *Environment International*, 32(8), 1033–42.

Wormuth, M., Scheringer, M., Vollenweider, M. and Hungerbühler, K. 2006. What are the Sources of Exposure to Eight Frequently Used Phthalic Acid Esters in Europeans? *Risk Analysis*, 26(3), 803–24.

Chapter 7

Sustainability in Agriculture and Food Production

Erling Jelsøe and Bente Kjærgård

Introduction

Agriculture and food production is one of the sectors that has been subject to most attention and debate regarding the question of sustainable development. The issue of food security was discussed in the Brundtland Report and seen as one of the most important challenges in relation to sustainability. It pointed to the paradox that more food than ever before is produced in the world today and yet hundreds of millions of people are starving. It also stressed that despite the considerable growth in food production that had been experienced since the Second World War, the signs of crises in food production were apparent all over the world. Thus, the achievements with respect to food production have been counterbalanced by the emergence of economic as well as ecological crises that were (and are) interlinked. The industrialised countries had growing problems with managing their food surpluses, the basis of income for small farmers in the developing countries was diminishing, and everywhere there were heavy pressures on the resource base.

Since then, a growing number of policy documents and programmes around the world from governments as well as international organisations have made sustainability a keyword in relation to the development of agriculture and food production. This holds for developed and developing countries alike, even though the issue of sustainable agriculture and food production is often associated with the situation in developing countries. Thus, in the EU, sustainability is mentioned as a key element of the Common Agricultural Policy (CAP), in Agenda 2000, and the subsequent reform of the CAP in 2003.

In the EU, environmental considerations have been increasingly incorporated in the CAP since the McSharry reform in 1992. This was a reaction to the growing problems of surplus production associated with the price mechanisms of the CAP, but it also reflected a growing recognition of the 'negative externalities' of the productivist paradigm (Lang and Heasman 2004). The main features of the productivist paradigm have been mechanisation, the use of fossil energy, growing chemical inputs in terms of synthetic fertiliser and pesticides, and large-scale production and specialisation as well as the industrialisation and concentration of animal farming. The negative externalities encompass the increasing consumption of energy, the pollution of groundwater and surface waters as a consequence of both the use of chemicals and handling of large amounts of manure, and changes in the countryside resulting in the reduction of biodiversity and social depletion. All in all, this development influenced sustainability negatively both in terms of environmental degradation and because of the social consequences for rural life, including the working conditions in agriculture.

In addition, the surpluses of food production together with the CAP trade regulations with export subsidies and tariff barriers represented a negative global relation between rich and poor countries to the detriment of farmers in the Third World.

Today, 20 years after the emergence of the Brundtland Report, we are still faced with many of these problems but the picture has become more complex. On the one hand a number of changes have occurred that have addressed the problems of the CAP and the negative externalities of the productivist paradigm. Thus, the 2003 reform of the CAP emphasises good environmental practice in farming and the development of rural areas. It has also diminished price support further and thereby curtailed the food surpluses. Trade relations with the surrounding world are also subject to change not least because of the influence of the World Trade Organization (WTO). Finally, the concept of multifunctionality in agriculture is often seen as a key to changes that may affect both the social and environmental components of sustainability in food production for the good.

On the other hand, many of the features of the productivist paradigm still prevail. Despite the changes in the CAP, the general traits of capitalist economy seem to favour ongoing centralisation and large-scale production as the dominant characteristics of agricultural development even though an orientation towards increasing interest in local produce and organic farming may counteract this to some extent, at least in some countries. Thus, this trend towards large-scale production and cost reduction must be seen as a phenomenon that is connected with the food chain as whole, that is, it prevails in all links of the chain from agriculture, the various links of industrial processing of food, to distribution and retail. Increasingly, large supermarket chains are setting the conditions for the development of all the links in the chain through detailed specifications of demands for price and quality of food products. In this context, it is still an open question as to what extent local markets for high quality foods will represent an alternative form of development. Furthermore, despite some improvements, the general picture of the condition of the countryside is still negative in environmental terms. The present situation, globally, with fluctuating food prices and trends towards use of biomass as a resource base for making biofuels on a large scale, fostered among other things by the present EU policy, may very well enhance the productivist character of agricultural production and increase the pressures on the natural environment.

So the present situation regarding the development of agriculture and food production, at least in the European context, seems to us to be one marked by contradictions rather than driven by the quest for sustainability as expressed through EU policies as well as in the many national policy papers that are the background for establishing programmes for sustainable development. The situation can perhaps also be characterised, as it is done by Drummond and Marsden (1999), as a paradox between a sustainability discourse and unsustainable material conditions.

Some of the reasons behind this paradox have to do with the framework for capitalist development of food and agricultural production. But the fact that sustainability is mentioned as a basic goal in so many policy programmes and decisions not least in relation to food and agriculture is in our minds an indication that the conceptual understanding of sustainability is part of the problem. Not in the simple sense that a debate and clarification of the concept of sustainability in itself is a solution, but in the sense that the understanding of what sustainability means is part of the social and political battlefield that surrounds food production. Thus the integrative character of the concept, cutting across social and natural conditions and the emphasis on both present and future needs is at the same time the great strength and significance of the concept and the basis for multiple social interpretations by different actors in society.

In this chapter we will therefore discuss the concept of sustainability in relation to food and agricultural production as a basis for discussing the fate of the concept as it is being used as a framework for the development of the sector. The first part of the chapter will be devoted to an examination of the way the concept is developed in different international contexts ranging from prominent institutional actors like the Food and Agriculture Organization (FAO), the Organization for Economic Cooperation and Development (OECD) and the EU to alternative conceptions launched by the International Federation of Organic Agriculture Movement (IFOAM), and more critical perspectives presented by individual contributions by various researchers. The aim is to demonstrate the scope of different understandings and through this to discuss some of the characteristic drawbacks or limitations.

Thus, we want to stress the importance of a broad, holistic concept of sustainability that takes into account all the dimensions inherent in the famous definition by the Brundtland Commission (1987), that sustainable development is 'a development that meets the needs of the present without compromising the ability of future generations to meet their own needs'. Contrary to this there is a tendency in political discourses on the issue to reduce the concept primarily to dealing with environmental aspects or even further to only selected dimensions of the environment (i.e. those that are most easy to handle politically), and to accentuate economic growth at the expense of social and cultural concerns.

The subsequent part of the chapter will highlight a specific example of a national political programme for sustainable development and in particular its section on food and agriculture. This is the Danish National Strategy for a Sustainable Development, which was launched in 2002 with later follow-ups. We will illustrate the use of the concept of sustainability in this programme and discuss some of its limitations. We will also discuss some of the global implications of the Danish programme even though our focus is mainly on the developed countries and their role in contributing to a sustainable development.

Finally, we want to introduce the need for what might be called 'food democracy' as a basis for realising a broad strategy towards sustainability since such a strategy inevitably has important implications for the way we live and may affect the dominant frames of globalisation.

Conceptualising Sustainable Agriculture and Food Production: Some Dilemmas

Balancing environmental, economic, social and institutional concerns is at the core of the concept of sustainable development. To illustrate the inherent ambiguities and dilemmas in conceptualising sustainable development we will take a critical look at some of the more powerful conceptualisations of sustainable agriculture. The idea is to open up a wider discussion on how to understand the sustainable dimensions of agriculture and food production, and to highlight some of the pitfalls or traps having dominated the debate about sustainable agriculture. We find that the proponents have a tendency to highlight the environmental dimensions and problems in comparison to social and cultural dimensions of sustainable agriculture. To illustrate the pitfalls or traps inherent in the global conceptualisations, we will take a closer look at how sustainable agriculture has been conceptualised and interpreted by international organisations/institutions and NGOs.

Environmental Sustainability as the Crux of Sustainable Agriculture

In 1995, the FAO conceptualised sustainable agriculture as:

> the management and conservation of the natural resource base, and the orientation of technological and institutional change to ensure the attainment and continued satisfaction of human needs for present and future generations. Such sustainable development in the agriculture, forestry, and fisheries sectors conserves land, water, plant and animal genetic resources and is environmentally non-degrading, technically appropriate, economically viable and socially acceptable. (FAO 1995, quoted from World Bank 2006: 2)

In many ways it is a definition involving all the dimensions to be expected, and hence it does not give much practical guidance as to what sustainable agriculture is really about or what it takes of institutional changes to move towards a more sustainable food production. Later, the FAO narrowed down the sustainable pathway of agriculture:

> increasingly to decouple agricultural intensification from environmental degradation through the exploitation of biological and ecological approaches to nutrient recycling, pest management and soil erosion control. (FAO 2003: 26)

By directing attention to the environmental dimension of sustainable agriculture and, in particular, strategies concerning how to cope with the environmental question, the prior holistic conceptualisation seems to have been abandoned or at least the environmental dimension has taken precedence over other dimensions such as the social one which is tantamount to focusing on new technologies and management strategies as the strategic measures to move towards sustainable agriculture.

Examples of reducing sustainable agriculture to a matter of a strategic focus on environmental sustainability can be found in the Agriculture Investment Sourcebook (World Bank 2006). The purpose of the Sourcebook is to support the implementation of the rural strategy 'Reaching the Rural Poor' by sharing information on investment options and innovative approaches as a basis for future lending programmes for agriculture. The Sourcebook singles out two areas for investments in sustainable agriculture: sustainable agricultural intensification and sustainable natural resource management. While the first is oriented towards on-farm investment, the latter focuses on off-farm investment. Sustainable agricultural intensification is prescribed as appropriate for improving the livelihood of farmers in developing countries and includes management strategies such as integrated pest management, conservation farming, low external input and sustainable agriculture, organic agriculture, precision agriculture and diversification. Drawing on the division of sustainable agriculture into environmental, social and economic sustainability, the Sourcebook rests on the understanding that:

> Environmental and social sustainability of productive resources depend in part on economic profitability that must provide for the maintenance of these resources (including the natural environment). (World Bank 2006: 4)

Though economic sustainability as well as environmental and social sustainability are seen as interdependent dimensions of sustainable agricultural strategies, it is apparent that sustainable agriculture is primarily addressed as a market-led environmental change furthered by public support or investments in new knowledge and information services and enabling public policies encouraging private investments in new technologies or management practices. It is evident from the Sourcebook that advocating for sustainable agricultural intensification as a livelihood strategy poses new dilemmas and problems as it is recommended to follow up investment in agricultural intensification and diversification with safeguarding policies such as environmental assessments of new agricultural production systems and pest management (World Bank 2006: 12).

While sustainable agricultural intensification often means a higher degree of relying on external resources/inputs, technologies and knowledge such as artificial fertilisers, pesticides, modern crop varieties, genetic modified seeds, in other cases sustainable intensification means emphasising the use of internal or local available resources and knowledge such as relying on regenerative and resource-conserving technologies and practices (Pretty 1995: 2). From the latter position sustainable agriculture is about making agriculture productive, environmentally sensitive and capable of preserving the social fabric of rural communities. Here the use of locally adapted resource conserving technologies has to be coordinated by groups or communities at the local level, and the role of external government or non-governmental institutions is working in partnerships with farmers (Pretty 1995: 3). From that standpoint, sustainable agricultural intensification means a reduction in external inputs and greater dependence on local resources and knowledge as the strategic focus for increasing yields. A better use of internal resources can be brought about '*by minimizing the external inputs used, by regenerating internal resources more effectively or by a combination of both*' (Pretty 1995: 9). A strategic focus on local inputs of resources and knowledge challenges the market-led environmental change of agriculture advocated by international institutions such as the World Bank and FAO. As the pressure for agricultural intensification is anticipated to increase towards 2030 similar to the increase in the preceding period, the FAO (2003: 23) argues that achieving sustainable agriculture will require trade-offs among the different dimensions of sustainability as the challenge is to keep agricultural production within limits that do not threaten the ability of future generations to achieve food security and acceptable effects on the wider environment. The remedy prescribed includes, among others, integrated pest management (biological, cultural, physical and when essential chemical pest management techniques), integrated plant nutrient systems, no-till/conservation agriculture and organic agriculture. Drawing on the decoupling of agricultural intensification and environmental degradation, the strategic focus is on greater use of biological and ecological approaches to nutrient recycling, pest management and soil erosion control (FAO 2003: 26).

Though the above-mentioned international institutions recognise that sustainable agriculture involves all dimensions of sustainability, it is the environmental dimension that dominates the discourse about sustainable agriculture. To illustrate the point further, we will give a few other examples of the precedence that environmental sustainability, and in particular market-led environmental change, has taken over other dimensions of sustainability in the debate about sustainable agriculture.

Like the above-mentioned international institutions, the OECD draws on an understanding of sustainable development as involving three pillars: economic, environmental and social sustainability (OECD 2006: 3) and in relation to agriculture emphasis is given to the linkages between the economic and environmental dimensions of sustainable development (OECD 2006: 35). While the multifunctional character of agriculture is outlined as a shared goal of the 1998 policy reform formulations of the OECD Committee for Agriculture at ministerial level (OECD 1998), it is abundantly clear that sustainable agriculture above all is linked to the environmental dimension as it is singled out as the priority area to:

> foster sustainable development through analysing and measuring the effects on the environment of domestic agricultural and agri-environmental policies and trade measures. (OECD 1998: section 17)

None of the other priority areas outlined explicitly mention sustainable development nor environmental sustainability or any of the other dimensions. It is obvious from the above quotes that sustainable agriculture is conceptualised in its narrow meaning. The narrow strategic focus is also reflected in more recent publications about the linkage between environment and agriculture (OECD 2001) and the linkages between policies, agriculture and environmental performance (OECD 2004).

The European Community draws on an understanding of sustainable development as about meeting the needs of the present generation without compromising the ability of future generations to meet their own needs' (COM 1992). Apart from environmental protection, social equity and cohesion, and economic prosperity, the Sustainable Development Strategy explicitly mentions international commitments (COM 2005). A closer look at a more recent policy document reveals that sustainable development is:

> about safeguarding the earth's capacity to support life in all its diversity. It is based on democracy, gender equality, solidarity, the rule of law and respect for fundamental rights. (European Council 2006: 7)

Though social and cultural dimensions of sustainability are explicitly mentioned, it seems that the environmental dimension takes precedence over other sustainability dimensions in the European Union policy formulations on agriculture. A clear signal of the strategic focus on environmental sustainability can be found in the Cardiff Meeting in June 1998 where it was decided to integrate environmental concerns into major policy proposals. The later 'Directions towards sustainable agriculture' (COM 1999) can be viewed as an offshoot of the Cardiff Meeting, though it was decided later in a follow-up meeting in Vienna. A closer look at the EU policy reform with regard to sustainable agriculture substantiates the strategic focus on environmental sustainability, although the policy formulations on the EU Commission's website 'Agriculture and Food' indicate a more comprehensive approach to sustainable agriculture as it is argued that achieving sustainable agriculture (in Europe) means meeting:

> an economic challenge (by strengthening the viability and competitiveness of the agricultural sector); a social challenge (by improving the living conditions and economic opportunities in rural areas); and an ecological challenge (by promoting good environmental practices as well as the provision of services linked to the maintenance of habitats, biodiversity and landscape). Sustainable agricultural production must also reflect the concerns of consumers, particularly as regards quality, safety and traditional/organic production methods. (COM 2006)

In 'Directions Towards Sustainable Agriculture' the pressure on the environment is attributed to the intensification of agriculture (COM 1999: 8). Here, the EU Commission puts sustainable agriculture into a different perspective and conceptualises sustainable agriculture from the self-interest of the farmers as 'a management of natural resources in a way which ensures that the benefits are also available in the future' (COM 1999: 6) while sustainable agriculture from a more comprehensive perspective is captured as 'the beneficial use of land and natural resources for agricultural production has also to be balanced with society's values relating to the protection of the environment and cultural heritage' (COM 1999: 6).

The early policy formulations towards sustainable agriculture do not leave much doubt that environmental sustainability is at the core of the conceptualisation of sustainable agriculture. A closer look at the policy reform formulations related to agriculture reveals that environmental considerations are to be subordinated to the CAP reforms orientation towards increasing the competitive position of the farmers (COM 1999: 22). Though it is explicitly formulated that 'agricultural activities should not pollute the environment, nor lead to serious erosion, nor destroy cultural landscapes'. It is also a clear priority that environmental requirements that 'go beyond good agricultural practice, which implies that the farmer already respects minimum environmental requirements' (COM 1999: 22) should be shifted to or compensated for by society. In fact it seems that increased competitiveness is given precedence over the environmental question in the EU agricultural policy.

Organic Agriculture as a Conceptualisation of Sustainable Agriculture

As an example of the challenges inherent in the conceptualisation of sustainable agriculture we will draw on IFOAM's conceptualisation of organic farming. Organic farming can be seen as one form of sustainable agriculture. The first IFOAM standards were published in 1980 and subject to biennial review (Kilcher, Huber and Schmid 2004). The basic IFOAM standards reflect the current state of organic practices around the world and are to be adapted to local conditions. The basic IFOAM standard provides a framework for standard setting, supervision and sanctions on a regional level. Though private standards have been influencing national and EU regulation on organic farming, and have been the basis for setting the IFOAM basic standards, the private standards no longer seems to have the upper hand as national regulations and authorities take over standard setting instead of farmers' associations such as is the case in Denmark, the EU and the US. Though standard setting seems to slip more and more out of the hands of the farmers' associations, IFOAM initiated a two-year consultative process concluding with four principles as a new basis for inspiring the future actions of the organic movement (IFOAM 2005). The four ethical principles embrace 'health', 'ecology', 'fairness' and 'care' and can be seen as an example of insisting on a more comprehensive approach for guiding global and local actions, adoption of technologies and practices of sustainable agriculture. Aiming at attaining ecological balance, genetic and agricultural diversity, the principles of health and ecology touch on scale, environmental quality and resource conservation as being basic for guiding agricultural practices. The principles of fairness and care aim at guiding relations and actions in the food chain through taking into consideration equity, justice and precautionary issues in the decision making about the environmental and social costs of the adopted technologies. According to IFOAM (2006), the four ethical principles adopted are to be turned into a 'succinct' definition of organic agriculture 'reflecting its true nature and the principles in a concise way'. The challenge when merging the four core principles into a concise definition of organic agriculture will be how to take into account variations in local conditions for farming in guiding actions for both policy reform and for the choice of appropriate technologies and management strategies. Whether it will be possible to define organic agriculture in a concise way is up for debate.

From our perspective, it seems a rather bold venture to aim at a concise definition of organic agriculture. New technologies, new knowledge about the environmental effects of current organic practices, or changes in social preferences towards fair trade are all changes that may open up the possibility for a change in the understanding of what organic agriculture is about and hence a future change in the definition of organic farming cannot be precluded. In fact, we believe that the principles set by IFOAM open the door for a continual discussion of what organic agriculture means and in continuation thereof what sustainability means in relation to organic agriculture. As the standard setting slips out of the hands of the farmers' associations, the role of the organic movement in defining sustainable agriculture is under pressure. With the states and intergovernmental institutions taking over the standard setting, the future challenge lies in driving improvements and innovation in organic practices and in guiding the political process related to organic standard setting. The four ethical principles are up against the mainstreaming of organic agriculture or the conventionalisation that have taken place since the late 1990s. As large-scale operations are not against the organic standards set, other drivers are needed to secure a visionary organic development on local and global levels. Though the ethical principle of fairness promises a refocusing on social equity issues, some authors point to the success of mainstreaming organic farming as the challenge to maintaining the commitment to the organic principles (Kristiansen and Merfield 2006).

Some Critical Perspectives on the Conceptualisation of Sustainable Agriculture

The above examples of the discourse on sustainable agriculture have been criticised from different perspectives. One of the criticisms is that it is a questionable venture to give a precise definition of sustainable agriculture. Pretty argues that 'sustainability itself is a complex and contested concept' and that discussions of sustainability need to clarify 'what is being sustained, for how long, for whose benefit and at whose cost, over what area and measured by what criteria. Answering these questions is difficult, as it means trading off values and beliefs' (1995: 11).

Apart from IFOAM's latest initiative, there is not much evidence that the above institutions will support a definition that will restrict the options available to farmers and local communities, i.e. by defining a concrete set of technologies, practices or policies as the core of the sustainable model. In fact it seems that the discourse is much more open, though much of the debate and the initiatives taken focus on environmental sustainability and in particular market-led environmental change of agriculture.

It seems that the challenge related to defining sustainable agriculture lies in identifying socially and environmentally acceptable trade-offs at the regional and international levels. For example Pretty (1995: 11) argues that it may be possible to trade off and agree on criteria for measuring trends in sustainability on farm and community levels, but the real challenge lies in employing such criteria for sustainable agriculture at the regional and international levels in a meaningful way. Others like Nielsen and Nielsen (2006: 265–76) point to the turn of the sustainability discourse towards a discourse on sustainable development and, in particular, the turn of the strategic meaning from a life-supportive function to securing a continual flow of natural resources for productive purposes. From their point of view, the turn means that a balanced use of natural resources is no longer an integrated part of sustainable development; instead the discourse is about controlling the use of natural resources and the ecological systems. In fact Nielsen and Nielsen (2006: 268–76) argue that the turn of the discourse not only highlights the ecological dimension but also reduces it to a question about wild nature. The vital questions about the use and control of biological resources and the subsequent consequences are left out of the discussions. In line with Sachs (1992), they criticise the concept of ecology and the concept of environment for being social constructions that reduce the understanding of nature to a self-regulating system aiming at calculating the assimilative capacity or adjusting feedback mechanism and hence reduce the natural world to an object for societal management (Nielsen and Nielsen 2006: 269–76). Though the authors are open to the possibility that the knowledge about the workings of ecological systems can be drafted and used differently, the critical perspective is clearly to overcome reducing the environmental question to an issue about imagining how ecological systems work and how that knowledge can be used for guiding ecological mainstreaming of industrialised agriculture or nature. The turn in the discourse also embraces a criticism of the division between environmental and social sustainability and hence an understanding of a divide between culture and nature. Nielsen and Nielsen (2006: 276) point out that with the turn, environmental sustainability seems to be displaced by economic sustainability as the integrative force of economic and ecological sustainability.

Inherent in the discourse about sustainable agriculture is what a sustainable relation between environmental, social and economic considerations in relation to agriculture means. The precedence given to the environmental dimension or more precisely a market-led environmental change of agriculture by international actors indicates that environmental sustainability is seen as a key to sustainable agriculture and as a goal compatible with a competitive and market-oriented agricultural policy. Targeting environmental issues is anticipated to benefit both farmers and society as it produces both public goods outcomes and economic growth. A central question for the debate about sustainable agriculture could be: What is the desired relation between agriculture and the environment? Is the intensification of agriculture as advocated by the international actors the key to sustainable agriculture? The conventionalisation of organic agriculture as is the case in Denmark is a good example of the dilemmas inherent in the intensification of agriculture. While large-scale organic operation may be perceived as a success on local and global levels if the success is measured in its income generation at farm level and the increase in organic supply on national scale and export volume. If measured in terms of its environmental benefits, large-scale organic dairy farming is not possible without a significant input of surplus manure from conventional farms and imported feeding stuff. While the economic benefits accrue to the organic farmers, others pay the environmental costs of the organic intensification. An extensive approach to large-scale organic operation could be a possible solution if access to adjoining land is an option. The question of intensification is a key issue, not just for organic farming but also for all kinds of sustainable agriculture. The intensification of agriculture is not only tangled or linked with the environmental issue, it embraces social and cultural issues such as the relationship between farmers, workers and consumers, animal welfare and ethical issues related to food additives or GMOs. From our perspective the future challenge lies in how to give social and cultural issues a more prestigious position in a market-led development towards sustainable agriculture.

A National Strategy for Sustainable Development

Intensification was (is) a dominant feature of the productivist paradigm, which was discussed in the introduction to this chapter. As mentioned, many of the features of this paradigm still prevail even though the situation characterising food and agricultural production since the beginning of the 1990s has become more complex. One interesting thing that has taken place is the loosening of the alliances and networks between agricultural organisations and the Government that supported the paradigm. Such networks and alliances have existed in many countries as well as on a European level in relation to the CAP (e.g. Daugbjerg 1998, 1999), and not least in countries with strong economic interests in food and agricultural exports such as Denmark. However, during the 1990s a number of policy initiatives related to the environmental impacts of agriculture and food production emerged and as we shall discuss further below some improvements in environmental performance also took place. This formed the context for the sections on food and agriculture of the Danish national strategy for sustainable development, which was launched in a number of Governmental Reports from 2001 onwards. The national strategy was set up in response to agreements of the UN Rio+5 Summit in 1997, and Denmark was one of the latecomers among EU countries in making such a strategy.

The first national strategy was made under the former Social Democratic Government in March 2001 (Danish Government 2001) and was soon substituted by a new one in June 2002 (Danish Government 2002) after the Liberal-Conservative Government had taken over later in 2001. Despite quite important differences in some respects regarding the environmental policy of the new Government, the new strategy differed only slightly from that of the Social Democratic Government, primarily in terms of a somewhat greater emphasis on the importance of economic growth. In the following we will therefore focus on the initiatives of the Liberal-Conservative Government. This strategy was presented internationally at the World Summit on sustainable development in Johannesburg in 2002, and a brief follow-up report after the Summit was published by the Danish Government in 2003 (Danish Government 2003). In 2007 the Government issued a discussion paper about the strategy for a sustainable development (Danish Government 2007).

Following the growing political emphasis on environmental aspects of food and agricultural production during the 1990s, the document presenting the strategy in 2002 expresses a rather clear support to what is conceived as goals of sustainable development:

> Denmark's objective for food production is to ensure that the food produced and sold to consumers is healthy and of high quality and that the level of information on food is high. Production methods that preserve the resource basis of the agricultural and fisheries sectors and secure the environment, nature, animal welfare and good working conditions must be promoted. Simultaneously, cost-effective production and marketing should be promoted in the food-producing sectors. Sustainable development of food production requires the right legislative framework, visionary utilisation and development of technological possibilities, and constructive interplay between the authorities, industry, and the public. (Danish Government 2002: 41)

To a certain extent all the components of the concept of sustainability, i.e. the environmental, the social and the economic, are included in this formulation of the goals for a sustainable development. There is also a reference to conditions of production that conserve the resource base and safeguard the environment and nature, so there is a clear relation to the Brundtland Commission's well-known definition of sustainability. The Danish Government also states that development since the middle of the 1990s has taken place along with a drop in the use of pesticides and nutrients. It stresses that the leaching of nitrogen has been reduced by 32 per cent from 1990 to 1999. It also mentions that the surplus of phosphorous in the soils has fallen by 30 per cent since 1985, and since 1994 more than 200 pesticides have been banned. Furthermore, there is a reference to the political initiatives, which purportedly have supported the development towards these improvements including action plans regarding the aquatic environment, pesticides and organic farming. There is also a reference to some of the areas, where the situation is still problematic, such as the need for further reduction of the evaporation of ammonia from livestock and above all the continued deterioration of the conditions for wild animals and plants and the decline in biodiversity. Social aspects are also discussed in connection with the Government's programme for rural development and the need for social development and employment in the rural areas, better working conditions and preservation of cultural values in the country. Furthermore, the document deals with consumer issues of food safety in Denmark as well as internationally, and it stresses the need for inspection and control as well as information and labelling and use of the precautionary principle regarding the regulation of pesticides. Health issues related to nutrition and obesity and a number of diseases connected with that are also covered and a continued effort is announced as a goal even though no specific measures are mentioned. All in all, however, it appears as if the claims of a development towards sustainability are justified. There is also, at least to some extent, a holistic approach in the sense that both the environmental and social problems of agriculture and problems that affect the consumer directly are included in the strategy.

However, reservations about this picture of the Danish situation and the objectives for a sustainable development can be made on at least two levels. The first one has to do with the factual circumstances, which are clearly illustrated by looking at what has happened since the strategy was launched in 2002. Thus the fall in the leaching of nitrogen has almost come to a halt, and in the midway evaluation of the Government's third action plan for the aquatic environment in Denmark in 2008 it was noted that there had been no significant reduction, and that the goals for a further reduction in the leaching of nitrogen until 2015 could not be reached through the measures of the action plan (Waagepetersen et al. 2008). Similarly, despite plans for the reduction in the use of pesticides in the period 2004–2009, there has been a growth in the use of pesticides, and new studies report about the occurrence of pesticides in the groundwater. Regarding biodiversity and the state of nature, it is clear that Denmark lags behind in its attempts to meet the EU goals of halting the loss of biodiversity in Europe in 2010. Furthermore, the Government has been criticised for not using all the EU funding that has been available for its Rural Development Programme. Finally, there are still problems with food safety, especially with respect to salmonella in pork, but these problems also reflect inadequate control with food quality due to cuts in Government expenditures, which led to the resignation of the Minister for consumer affairs after political criticism in 2006.

The Government's 2007 discussion paper did not mention these problems with the fulfilment of the goals, which may be due to the fact that the evaluations of the Government's plans and efforts had not been finished by the time of the paper's publication. Not surprisingly, it paid much attention to climate issues, but apart from that it was remarkable that this new discussion paper had a much more one-sided focus on environmental sustainability than the 2002 strategy. Social aspects like the development of the rural areas or consumer issues related to food safety and health were not included. In that respect, the paper represented a step backwards in the conceptual understanding of sustainable development.

We will not go into a detailed discussion and explanation of these apparent inconsistencies in the way the Danish Government has handled the approach to the sustainable development of agriculture and food production. In general terms it illustrates the contradictions in the present situation, in which sustainability is a keyword for the development of agriculture and food production in policy documents and programmes, not least in the EU, whereas many of the problems of the productivist paradigm prevail. On the other hand, it must be recognised that much has happened both at the policy level and in some respects also regarding the actual state of the environment since the 1980s. This seems to be the case in most EU countries although the situation varies substantially according to differences in structure and the economic significance of food production, historical and cultural traditions as well as in climate and natural conditions. But despite the changes that have taken place over the last 20 years, there is still what Drummond and Marsden call a paradox between the sustainability discourse and unsustainable material conditions.

The paradox can be understood by considering the dynamics of the development of the food and agricultural sector. The Danish Government seemed to be aware of this already in its first strategic document published in 2002, where it pointed to the increased intensification as a result of specialisation and large-scale farming and the subsequent reduction in the numbers of farms, so that nearly 90 per cent of agricultural production today takes place on about 23,400 farms operated full-time.[1] It noted that intensive farming and the size of the fields reduces biodiversity and acknowledged that:

> sustainable agricultural production poses a range of dilemmas. At the same time as taking account of profitable production, the environment, biodiversity, health and safety, animal welfare, landscape values and rural development must also be considered. These concerns may well be conflicting. (Danish Government 2002: 45)

The way out of these dilemmas for the Danish Government has been found in the priorities set, which are reflected in the discrepancies we have discussed above.

---

1   This was the number of full-time farms in Denmark in 1999. In 2007 the number of full-time farms had gone down to 14,500, whereas in 1973 the number of full-time farms was 81,600. The average size of a full-time farm was 29.4 ha in 1973 and 131.4 ha in 2007 (Institute of Food and Resource Economics 2009).

However, these dilemmas do not just give rise to contradictions between the policy programmes and actual development. They also shape the more strategic approach to the development of food and agriculture. This is quite clearly illustrated through the most recent policy initiative from the Danish Government called 'Green growth' (April 2009). This is a plan for Danish agriculture, which tries to unify goals of economic growth and green solutions. It presents many of the goals from earlier plans, e.g. regarding reductions of the leaching of nitrogen and the use of the most problematic pesticides as well as the development of the rural areas and improvement of food safety. The new feature of the plan is its emphasis on growth through the removal of the hitherto existing restrictions on the number of animals that were allowed within a given farm area and on limits to the maximal size of a farm and animals. In addition, it contains a proposal for putting an end to the traditional principle of freehold by the farmer of his farm (i.e. the family farm), so that farms can be owned by a company. This opens up the possibility for new large-scale and 'landless' factory farms in the future. The prerequisite is that the farm must be able to handle the manure in an environmentally safe way, and it stipulates stronger demands concerning the documentation of compliance with regulation of environment and food safety. It launches a strategy for the handling of manure through the production of biogas and incineration of the solid components of the manure, which if realised will lead to the production of 'green energy' from 40 per cent of all manure from Danish agriculture and thereby a reduction in the annual CO_2 emissions in Denmark by 180,000 tons.

Taken at its face value this appears to be a plan for ecological modernisation of Danish agriculture and food production that combines increased competitiveness with the solving of environmental problems. This is also evident from the fact that it emphasises competitiveness through innovation in both agriculture and the food industry ('Denmark as a green laboratory for growth'). Furthermore, organic farming will be promoted not least with the aim of increasing exports of organic foods. It may be disputed to what extent this picture holds, as it has been pointed out that many of the Government's environmental goals are not very ambitious, and it turns out that the incineration of manure without biogas production may be the dominant scenario. Incineration will be associated with loss of nutrients and a reduction of the carbon pool in the soil. The support of organic farming with an emphasis on innovation and exports is likely to enhance conventionalisation as discussed above.

Some authors have pointed out that conceptualising the ongoing development of agriculture in Europe in terms of ecological modernisation is more adequate than talking about a transition to post-productivism (Evans, Morris and Winther 2002). We agree in this in the sense that despite the changes that have taken place over the last 15–20 years there is still an emphasis on production, intensification and large-scale operation as the recent Danish plan for green growth as well as empirical evidence regarding the current development of agriculture and food production clearly demonstrates. However, the Danish development also suggests that while the actual plan is indeed a programme for further modernisation it is more doubtful to what extent it is ecologically sound.

Some Global and Cross-sectoral Issues

In addition, there are issues that are not even mentioned in the Danish programmes for sustainability. Characteristically, these have to do with problems of a wider, even global character, or issues that involve larger parts of the food chains including lifestyles and food habits. Thus, the development of the food sector is also characterised by a large consumption of energy. This is not only the case because of the mechanisation of all operations within modern intensive farming systems and the energy content of inputs like fertilisers and pesticides. The capital concentration in all links of the food chain has, together with the internationalisation of production, led to the increasing transport of food products from producer to consumer and consequently an increase in the consumption of fossil fuels. The *food miles* issue has been addressed by Pretty et al. (2005) who show that the transport of agricultural and food produce is increasing and accounts for 28 per cent of goods transported on UK roads. There is no reason to believe that the situation in that respect should be much different in Denmark or other developed countries. Externalities include environmental, social and health problems related to traffic congestion, harm to health (air pollution and noise), greenhouse gas emissions and infrastructure damage. The air transport of foods globally is currently of minor importance in this connection because the total volume is still low, but it is increasing and potentially a severe problem.

In addition, there is the road transport of foods by consumers from shops to home. This, too, is increasing and some estimates indicate that this is equal to or even larger than the transport of foods from food processors to retailers (Garnett 2003, Jespersen 2004). A number of factors influence this development such as the emergence of large shopping centres, the location of the shopping centres out of or on the edge of towns and centralisation in the retail sector in general.

Another important issue is meat consumption, which is increasing globally and according to current estimates it is likely to double from now to 2050. The rapidly growing livestock production represents maybe the most serious problem in relation to the environment and resource depletion in the global food production. As stated in a major study published by FAO:

> Livestock's impact on the environment is already huge, and it is growing and rapidly changing. Global demand for meat, milk and eggs is fast increasing, driven by rising incomes, growing populations and urbanisation. (Steinfeld et al. 2006)

Thus, increasing meat consumption is intimately connected with major societal changes that take place almost globally with the largest increases taking place in the growth centres in East and South Asia. By far the highest consumption of meat is still found in Europe and the United States.[2] It affects most aspects of the environment like greenhouse gas emissions, soil erosion, water resources and quality as well as biodiversity. It also has wider implications for land use such as the increasing production of feed. This is indeed a global issue as, for instance, 42 per cent of all protein fed to Danish domestic animals came from imported feedstuffs, not least from soybeans from Latin America. Globally, the production of soybeans for animal feed has more than doubled since 1990.

It has been estimated that food consumption patterns, and in particular meat consumption, will be just as important and maybe even more important for the land required for food production as population growth (Gerbens-Leenes and Nonhebel 2001). The developed countries are still responsible for a large part of the world's meat consumption, and meat consumption in these countries also affects the rest of the world because of the pressure on natural resources through imports of feedstuffs and meat.

Similar problems exist in relation to the attempts to increase the production of biofuels that compete with food production regarding the use of agricultural areas and can put pressure on food production in the direction of further intensification and unsustainable ways of farming. Likewise, considerable doubt has been cast as to the sustainability of the production of a number of the crops for biofuel. The EU is also responsible in this connection because its attempts to increase the use of biofuels in European cars and it is still unclear to what extent the current attempts in the EU and some European countries to develop certification schemes and sustainable criteria for biofuel production will lead to sustainable solutions.

These questions show the complexity that surrounds the development of sustainable food production and that to a large extent – and in particular in relation to some of the more difficult issues – it involves global relations and cuts across different sectors such as food, energy and transport, and vertically through the food chains. In that sense, these issues raise challenges and dilemmas that a national government cannot cope with on its own but it is remarkable the way the Danish Government to a large extent ignores them in its attempts to reconcile the different and conflicting objectives.

Social Conditions for Change – Food Democracy

The most obvious implication of the way we see the current development of the food sector is that sustainability will not be a result of an almost quasi-

---

2 The average annual meat consumption per capita in developed countries was 78 kg in 2002, whereas for developing countries it was only 28 kg. However, the annual consumption per capita in developing countries has doubled since 1980. The Danish consumption of meat is even higher than average for developed countries, reaching 114 kg in 2002 (FAO 2006).

automatic ongoing trend towards a 'greening' of the food sector. The question of sustainability implies important and fundamental choices regarding the future development of food production and some of the barriers to change are intimately connected with dominant trends in current developments like globalisation, the dominance of large supermarket chains, etc. Currently, large supermarket chains like Wal-Mart are attracting attention through attempts to develop their green profile. Without going into a closer discussion about whether this represents a step towards sustainability or rather new attempts of 'greenwash' (e.g. Wal-Mart Watch 2007), there are, just as with governmental plans such as those discussed above, good reasons not to take for granted that such market-oriented approaches by themselves lead to a sustainable development of food production, given the complexity and challenges that are associated with such a development. Furthermore, supermarkets like governments have agendas of their own. Thus, Wilkinson (2006) concludes on the basis of his analysis of the global value chain of fish production (which we have delimited ourselves from discussing in this chapter) that the dilemmas and problems of the global value chain of food production and consumption 'cannot be left to the supermarket's putative capacity to internalise consumer demand' but must involve the 'consumer-citizen' more directly in a complex mix of direct action, new consumer practices, negotiated initiatives with industry and pressure for public, national and global regulation. As an example, he mentions how supermarket chains resisted proposals for informing consumers about whether fish products originated from aquaculture or not, because of fear that consumers would be less positive towards farmed fish.

What Wilkinson calls for, in our interpretation, is a broader involvement of citizens in their role as consumers or citizens or both in the development of food production towards sustainable development. Clearly such an involvement does exist in the form of social movements reacting against the productivist development of the food system. The organic farming movement, represented among others by IFOAM as mentioned above, is probably one of the oldest and most prominent examples and it still prevails, even though it is confronted with some policy dilemmas because of the integration of the organic sector into the global market economy. The slow food movement has also created considerable international attention and activities as a proponent for an alternative food culture, which is both more sustainable and revitalises a local and authentic food culture as an alternative to the internationalised fast food culture. Finally, there have been attempts to promote local foods at many places both in Europe and the United States.

Exactly how strong these trends are is difficult to assess, but despite the relative success of organic farming in many countries, these developments all appear so far to be weak and for the time being only marginal in the overall development of the food sector. On the other hand, it is clear and also recognised by more established political decision-makers that there is a need for a broader involvement of the population in promoting sustainable development. Such a recognition is also found in the Brundtland Report and can also be seen in the Danish Government's 2002 programme that refers to the Brundtland Report as well as to the Aarhus Convention about the involvement of the citizens in the environmental area and to the Danish tradition for citizen involvement. Apart from references to the already existing activities related to Local Agenda 21 (LA21), concrete objectives or proposals for initiatives in the field were relatively weak, however.

All in all, there are as indicated above very good reasons to discuss the question about a broader involvement of citizens in processes of change in the food system. In this context we will briefly mention two different contributions with a somewhat different emphasis to a discussion about this issue.

Lang and Heasman (2004) introduce the concept of 'food democracy' as opposed to 'food control' and by doing so, the need to involve the population more broadly in the development of food production and supply. Even though their book is mainly concerned with the health aspects of food, their approach is relevant to a discussion about sustainability as well. It is oriented towards more democratic forms of governance that involve a multiplicity of popular organisations and interest groups and maybe in particular to 'full stakeholder participation giving priority to including voices previously excluded'. They also emphasise that such a food democracy must engage with the real complexity of multi-level policy and control.

Alternatively, Kloppenburg et al. (2000) discuss an approach based on their experience regarding the involvement of 'competent, ordinary people' in an exercise in which the participants, who were a diverse group of people with an interest in 'the alternative food/farm community', were asked to present their visions of a sustainable food system. The exercise resulted in a broad range of ideas that exceeded the conceptual approaches found in an otherwise diverse range of documents made by professional groups and organisations or by academics. This exercise thus confirmed the experience that involving ordinary people in the construction of visions for change can be an important source of new ideas. The initiative was also interesting because it involved a combination of a broad range of people from farmers to marketers of food and consumers.

We are not advocating for one of these approaches at the expense of the other. It is our contention that both will be important in a development towards sustainability of the food system, that is, new types of democratic governance as well as various forms of participatory exercises that can invoke new ideas and proposals for change.

Conclusion

The complexity and conflicting interests regarding sustainability are clearly expressed in the many different attempts to conceptualise sustainability. The precedence of environmental sustainability is demonstrated in most of the documents from the international institutions despite references to the social dimension. The tendency is that the three dimensions – the environmental, the social and the economic – typically are mentioned in general discussions about the concept of sustainability, whereas when the more strategic priorities are at stake it is the environmental and economic dimensions that dominate, and the social will often both as a goal in itself and as an integrated part of the conception of the food system as a totality of nature and human activity be downgraded or marginalised.

Our analysis of the Danish programmes and policy documents reveals the same tendency at a more concrete level, where the consequences of the programmes become clear in relation to the actual state of the food system and its natural and social basis. Along with the development of the programmes during the last six to eight years they have increasingly turned into political support for a continued productivist development albeit with an environmental component that was almost absent before 1990. However, it can still be disputed to what extent they meet the environmental objectives, for instance regarding the aquatic environment and the reduction of biodiversity. The lack of an explicit discussion of global and cross-sectoral relations of the food systems further demonstrates the problems and points to the implicit structural constraint, which these relations seem to inflict on strategic thinking. New democratic forms of governance and more involvement of the population in formulating visions and ideas to promote change may open up the possibility for new social dynamics and even though the present constraints will still represent a great challenge it can relate to the albeit contradictory but still emerging social and environmental commitments, which also exist in and around the food systems.

References

COM. 1992. *5th Environmental Action Programme: Towards Sustainability.* 23 Final. Brussels.

COM. 1999. *Directions Towards Sustainable Agriculture.* European Commission communication. 22 Final. Brussels.

COM. 2005. *On the Review of the Sustainable Development Strategy: A Platform for Action.* Communication from the Commission to the Council and the European Parliament. 658 Final. Brussels.

COM. 2006. *Presidency Conclusions of the Brussels European Council* (15/16 June 2006). Brussels: Council of the European Union.

Danish Government. 2001. *Forslag til Danmarks strategi for bæredygtig udvikling. Udvikling med omtanke – fælles ansvar.* Copenhagen: Environmental Protection Agency. (In Danish).

Danish Government. 2002. *Denmark's National Strategy for Sustainable Development. A Shared Future: Balanced Development.* Copenhagen: Environmental Protection Agency.

Danish Government. 2003. *The World Summit in Johannesburg and Denmark's National Strategy for Sustainable Development.* Copenhagen: Environmental Protection Agency.

Danish Government. 2007. *Grønt ansvar. Regeringens debatoplæg om en strategi for en bæredygtig udvikling.* Copenhagen: Ministry of the Environment. (In Danish).

Danish Government. 2009. *Grøn vækst.* Copenhagen: Ministry of Finance. (In Danish).

Daugbjerg, C. 1998. *Policy Networks Under Pressure: Pollution Control, Policy Reform and The Power of Farmers.* Aldershot: Ashgate.

Daugbjerg, C. 1999. Reforming the CAP: Policy Networks and Broader Institutional Change. *Journal of Common Market Studies*, 37(3), 407–28.

Drummond, I. and Marsden, T. 1999. *The Condition of Sustainability.* London and New York: Routledge.

EU COM. 2006. *Agriculture and Food.* [Online]. Available at: http://ec.europa.eu/agriculture/foodqual/sustain_en.htm [accessed: 4 December 2006].

Evans, N., Morris C. and Winter, M. 2002. Conceptualising Agriculture: A Critique of Post-productivism as the New Orthodoxy. *Progress in Human Geography*, 26(3), 313–32.

FAO 2003. *World Agriculture: Towards 2015/2030. An FAO Perspective*, edited by Jelle Bruinsma. London: Earthscan.

FAO 2006. FAO statistical databases. [Online]. Available at: http://faostat.fao.org/defaultu.aspx [accessed: 11 October 2009].

FAO. 2006. *Agriculture Investment Sourcebook, Module 4: Investments in Sustainable Agricultural Intensification. World Bank 2006.* [Online]. Available at: http://www.worldbank.org/agssourcebook [accessed: 4 December 2006].

Garnett, T. 2003. *Wise Moves: Exploring the Relationship between Food, Transport and CO_2.* London: Transport 2000 Trust [Online]. Available at: http://www.transport2000.org.uk [accessed: 9 November 2009].

Gerbens-Leenes, P.W. and Nonhebel, S. 2002. Consumption Patterns and their Effects on Land Required for Food. *Ecological Economics*, 42, 185–99.

IFOAM. 2005. *Principles of Organic Agriculture: Preamble.* [Online]. Available at: http://www.ifoam.org/about_ifoam/principles/index.html [accessed: 4 December 2006].

IFOAM. 2006. *Definition of Organic Agriculture. Consultative Process: Work Plan.* [Online]. Available at: http://www.ifoam.org/organic_facts/doa/index.html [accessed: 24 July 2007].

Institute of Food and Resource Economics. 2009. *Statistics.* [Online]. Available at: http://www.foi.life.ku.dk/English/Statistics/Agriculture.aspx [accessed: 7 October 2009].

Jespersen, P.H. 2004. *The Transport Content of Products. Worlds Transport Policy and Practice.* [Online] 10(3) 28–35. Available at: http://www.ecologica.co.uk/wtpp10.3.pdf [accessed: 9 November 2009].

Kilcher, L., Huber, B. and Schmid, O. 2004. Standards and Regulations, in *The World of Organic Agriculture. Statistics and Emerging Trend*, edited by H. Willer and M. Yussefi. Bonn: IFOAM.

Kloppenburg, J., Lezberg, S., De Master, K., Stevenson, G.W. and Hendrickson, J. 2000. Tasting Food, Tasting Sustainability: Defining the Attributes of an Alternative Food System with Competent Ordinary People. *Human Organization*, 59(2), 177–86.

Kristiansen, P. and Merfield, C. 2006. Overview of Organic Agriculture, in *Organic Agriculture – A Global Perspective*, edited by P. Kristiansen, A. Taji and J. Reganold. Collingwood: CSIRO Publishing.

Lang, T. and Heasman, M. 2004. *Food Wars. The Global Battle for Mouths, Minds and Markets*. London; Stirling, VA: Earthscan.

Nielsen, B.S. and Nielsen, K.A. 2006. *En menneskelig natur: Aktionsforskning for bæredygtighed og politisk kultur.* Copenhagen: Frydenlund.

OECD. 1998. *Ministerial Communiqués Related to Agricultural Policies.* OECD Committee for Agriculture at Ministerial Level, March 1998. [Online]. Available at: http://www.oecd.org/document/34/0,3343,en_2649_33773_31852962_1_11_1,00.html [accessed: 23 July 2007].

OECD. 2001. *Adopting Technologies for Sustainable Farming Systems: Wageningen Workshop Procedings.* [Online]. Available at: http://www.oecd.org/dataoecd/40/26/2739771.pdf [accessed: 23 July 2007].

OECD. 2004. *Agriculture and the Environment: Lessons Learned from a Decade of OECD Work.* Joint Working Party on Agriculture and the Environment. [Online]. Available at: http://www.oecd.org/dataoecd/15/28/33913449.pdf [accessed: 23 July 2007].

OECD. 2006. *2006 Annual Report on Sustainable Development Work in the OECD.* [Online]. Available at: http://www.oecd.org/dataoecd/57/62/38392143.pdf [accessed: 20 July 2007].

Pétry, F. 1995. *Sustainability Issues in Agricultural and Rural Development Policies*. Rome: FAO.

Pretty, J.N. 1995. *Regenerating Agriculture: Policies and Practice for Sustainability and Self-reliance.* London: Earthscan.

Pretty, J.N., Ball, A.S., Lang, T. and Morison, J.I.L. 2005. Farm Costs and Food Miles: An Assessment of the Full Costs of the UK Weekly Food Basket. *Food Policy*, 30(1), 1–19.

Steinfeld, H., Gerber, P., Wassenaar, T., Castel, V., Rosales, M. and de Haan, C. 2006. *Livestock's Long Shadow: Environmental Issues and Options*. Rome: FAO.

Waagepetersen, J., Grant, R., Børgesen, C.D. and Iversen, T.M. 2008. *Midtvejsevaluering af vandmiljøplan III.* Aarhus: Aarhus University.

Wal-Mart Watch. 2007. *It's Not Easy Being Green: The Truth Behind Wal-Mart's Environmental Makeover.* [Online]. Available at: http://walmartwatch.com/pages/recent_Reports [accessed: 11 October 2009].

Wilkinson, J. 2006. Fish: A Global Value Chain Driven onto the Rocks. *Sociologia Ruralis*, 46(2), 139–53.

Willer, H. and Yussefi, M. (eds) 2004. *The World of Organic Agriculture. Statistics and Emerging Trends.* Bonn: IFOAM.

World Bank 2006. *Agriculture Investment Sourcebook. Module 4: Investments in Sustainable Agricultural Intensification.* [Online]. Available at: http://www.worldbank.org/agssourcebook [accessed: 4 December 2006].

World Commission on Environment and Development 1987. *Our Common Future.* Oxford: Oxford University Press.

Chapter 8

Sustainable Development, Capabilities and the Missing Case of Mental Health[1]

Andrew Crabtree

> What needs to be conserved are the *opportunities* of future generations to lead worthwhile lives. The fact of sustainability (in both production and consumption) implies that what we are obligated to leave behind is a generalized capacity to create well-being, not any particular thing or any particular resource. Since we do not know what the tastes and preferences of future generations will be, and what they will do, we can talk of sustainability only in terms of conserving a capacity to produce well-being. (Anand and Sen 2000: 2038) [Emphasis in the original]

> It is true that we do not know what the precise tastes of our remote descendants will be, but they are unlikely to include a desire for skin cancer, soil erosion, or the inundation of all low-lying areas as a result of the melting of the ice-caps. And, other things being equal, the interests of future generations cannot be harmed by our leaving them more choices than fewer. (Barry 1991: 248)

Definitions of sustainable development now abound, but that offered by the Brundtland Commission is still the most commonly cited and, as made clear in the introduction to this book, it continues to provide the overarching framework for discussions about sustainability, hence it is the point of departure here. To quote:

> Humanity has the ability to make development sustainable – to make sure that it meets the needs of the present without compromising the ability of future generations to meet their own needs. (World Commission on Environment and Development 1987: 8)

The Commission's idea was not that the world should be sustained as it was (or is) as there are billions of people in the world whose basic needs however loosely defined are not met – a fact that is reflected by the Millennium Development Goals and that, for the first time in *human* history, there are over 1 billion undernourished people. The Commission did not support the idea that growth was all there is to

---

1   I am indebted to many for their comments on earlier drafts of this chapter, but the most penetrating ones, for which I am most grateful, came from the now, sadly, late Professor Tom Whiston.

the story, growth is important (no one was arguing that stagnation or recessions were desirable or are good for the poor, they are not, as the present financial crisis attests),[2] but the Commission maintained that there is also a moral obligation to meet people's basic needs. Furthermore, it was argued that there are environmental constraints on the kind of growth permissible.

Whilst meeting basic needs was seen as paramount, the Report also included the less discussed concept of 'legitimate aspirations', for example:

> The satisfaction of human needs and aspirations is the major objective of development. The essential needs – for food, clothing, shelter, jobs – are not being met, and beyond their basic needs these [sic] people have legitimate aspirations for an improved quality of life. (World Commission on Environment and Development 1987)

In other words, development does not end when we have met everyone's basic needs.

Thus the Brundtland concept of sustainable development has three overriding themes: sustainability (limitations), development (basic needs and legitimate aspirations) and a view of ethics in terms of our obligations to others of both present and future generations to meet their basic needs and legitimate aspirations.

This chapter concentrates on just two of these aspects, the notions of development and ethics, and it contends that the Brundtland approach or indeed a needs approach per se is fundamentally flawed, and argues instead for a capabilities approach to sustainable development inspired by the works of the Indian Nobel Prize winning economist and philosopher Amartya Sen. For Sen, development is not to be seen as industrialisation or modernisation or the straightforward increase in gross national product (GNP) per capita or, one could add, although Sen does not discuss it, ecological modernisation. Rather, it is about increasing people's capabilities, the real freedoms people have to lead the lives they value (Sen 1999).

This chapter replaces the idea of basic needs with one of *essential capabilities*, both individual (bodily health, agency and social inclusion) and collective (production, reproduction and social capital). Such capabilities are those which are *universally necessary for present and coming generations to lead the lives they value*, and the concept of 'legitimate aspirations' with one of 'legitimate freedoms'. As to which freedoms are legitimate is an ethical question, and it is argued here that the answers to such questions can best be found in a Scanlonion contractualist approach to ethics and rights (Scanlon 1998, Barry 1995). The basic idea in the Scanlonian approach is that an action (I would add, a lifestyle) is morally wrong

---

2   In terms of carbon emissions, however, the picture looks rosier. The International Energy Agency has estimated that they have been reduced by 3 per cent. Available at: http://www.worldenergyoutlook.org/2009_excerpt.asp [accessed 15 October 2009]. Needless to say, there are other ways of reducing carbon emissions.

if the actor cannot justify his or her action to others on grounds that they could not reasonably reject. Thus the notion of legitimate freedoms developed here refers to *freedoms to live valued lives that cannot be reasonably rejected by others*. At heart, the capabilities approach endorsed here is emancipatory and entails substantial societal change.

This re-conceptualisation of sustainable development also leads to changes in policy focus. The specific issue that is taken up here is that of mental health. The essential capability of agency draws our attention to mental disorders which are one form of limit on agency (others would, for example, include social structures or lack of physical mobility (see Figueroa, this volume)). Until now, mental health issues have remained peripheral to the sustainability debate, for instance, the UN sustainable development indicators only include suicide for mental health, and their data is incorrect.[3] The United Kingdom sustainable development indicators have now added one for positive mental health (Warwick Edinburgh Well Being Scale) stating 'Measuring mental health from a positive perspective would be a significant development from the traditional measurement of negative mental health or illness' (DEFRA 2007: 131). Positive mental health is clearly significant, yet it is not apparent why negative mental health should be left out of the picture, not accounting for its magnitude, severity or diversity. Surely, positive mental health should be added to the list rather than replacing the information we have about mental disorders. If indicators of sustainable development are meant to guide policy, it is a strange move to ignore the vital information we already have and policy makers already use.

The chapter develops as follows: the next section outlines the concept of basic needs and the reasons for rejecting it. Thereafter, Sen's capabilities approach is introduced, its links to Scanlonian contractualism explored, and the notion of legitimate freedoms developed. I then go on to elucidate my concept of essential capabilities, which leads to a discussion of the importance of mental disorders and a conceptualisation of their relationship with the environment. The chapter then concludes.

Brundtland and Needs

As many chapters in this volume attest, the notion of basic needs has been central to the sustainability debate, nevertheless the concept remains unclear. What is a *basic* need and which needs should be met? The answers to these questions are important for guiding policy objectives, yet the Brundtland Report remains unclear on this account, it gives us examples of the needs that might be included, and in *some* cases we are given justifications. However, lists vary. In addition to food, clothing, shelter and jobs listed in the quote above, we also find energy, water supply, sanitation and health care (World Commission on Environment and

---

3   They state that there are 100,000 suicides a year, the World Health Organization's (WHO 2001a) figure is 1 million.

Development 1987: 55). Of these, employment is seen as the most basic of all needs. The concept of employment is qualified by bringing in the notions of time and choice 'sustainable work opportunities', and given an end 'to meet minimum consumption standards' (World Commission on Environment and Development 1987: 54), though again these notions are not defined. The variation in lists makes it difficult to establish which needs are basic, on what criteria, as to whether we can add needs at will, or as to whether the list should be left open and, if so, why. The Commission leaves all these central issues unresolved. Furthermore, it does not provide us with an ethical justification for our obligations to meet such needs.

Can the concept of sustainable development be saved by appealing to more sophisticated expositions of the notion of needs and justifications for meeting them? After all, the Report is a report and not a scientific or philosophical treatise. Could one not take the Report's basic idea and provide a stout defence of the notion of needs making up for the Report's weaknesses but maintaining the Report's thrust? Attempts to provide a rigorous notion of needs exist (in particular Wiggins 1987, Doyal and Gough 1991). Although they give sustainable development a side role, one could attempt to build on them.

This path is rejected here for three fundamental reasons. Firstly, in so far as the needs approach concentrates on resources (food, housing, sanitation, etc.) as the Brundtland Report to a large extent does, it runs into the conversion problem, namely that different people require different resources to achieve the same objectives (Sen 1999). Some people need wheelchairs, ramps, special buses, etc. to get about whereas others do not. Food requirements differ significantly from babies to lactating mothers to working men to old people and so on or, to use one of Nussbaum's examples, in countries where female education is opposed, more resources will probably be needed to ensure educational equality (Nussbaum 2004). Accordingly, we should concentrate on what people are actually able to do or to be, i.e. their capabilities (to get around or to be adequately nourished, to be educated) rather than the resources that are necessary to meet those ends. Secondly, resources are not enough in the sense that their proper use requires an understanding of how to use them, thus there are cases where, for example, water sanitation resources have been provided but not used. Thirdly, basic needs approaches whether they are resource based or not do not make choice central to their account. For example, Doyal and Gough, who do not endorse a resource-based view, argue that health is a basic need. Their suggested indicators include life expectancy, age-specific mortality rates, the prevalence of disabilities, children suffering from developmental deficiencies, people suffering from serious pain and 'morbidity for various disease categories' (Doyal and Gough 1991: 190).

Anyone who falls under that line is not having their basic needs met, it is society's duty to do so. However, such an approach ignores the real opportunities (capabilities) people have. For example, some people die earlier than they otherwise would due to various lifestyle diseases such as cancer caused by smoking or, again, the calorie intake of someone who is starving during a famine may be the same as that of a so-called supermodel and be below 2,100 calories a day, however, they have different capabilities, namely the supermodel could choose to eat more but decides to diet, whereas someone in a famine does not have that choice, and our obligations are correspondingly different. Consequently it is people's real opportunities rather than people's states that are important, and thus provide the appropriate evaluative space.

The Capabilities Approach and Scanlonian Contractualism

The capabilities approach was first introduced by Amartya Sen in his Tanner Lecture on human values in 1979 (Sen 1980). Since then the approach has gained widespread influence in the development field not least because of Sen's influence on the Human Development Reports which are the major alternative to the income/growth approach propagated by the International Monetary Fund and the World Bank (now modified by the Millennium Development Goals). The approach has also been developed within political philosophy by Martha Nussbaum (2001, 2006), and while there are different goals and interests there is considerable interface between the two strands. However, I will concentrate on Sen's version here (for an introduction to the approach more generally, see Robeyns 2005).

As already stated, for Sen, capabilities are the real freedoms that people have to lead the lives they value, that is, to attain certain functionings. 'Real freedoms' differ from 'freedoms' in that people can exercise them. For example, according to the Indian constitution, schedule castes have the same rights as all other Indians, or in some cases receive positive discrimination, but in practice these are often flouted and when so, scheduled castes' *real* freedoms are limited. Functionings refers to actual doings and beings, for example being healthy or educated, swimming, eating ice cream or playing the blues. The relationship between real freedoms and functions can be explained by using one of Ingrid Robeyns's examples: a battered boxer and a battered wife may have the same functionings (being battered) but have different capability sets – the boxer chooses to enter the ring, the wife, we may presume, does not.

Of course, not all the lives people value living are acceptable, and it would be wrong to understand Sen as endorsing such a view (see Sen 2004, Anand and Sen 2000) even if his major expositions of the capabilities approach do not say so, and appear to endorse freedoms *ad libitum* (e.g. Sen 1999) leading to accusations of 'freedom fetishism' (Reader 2006). This issue is particularly relevant in relationship to sustainable development, as some freedoms and lifestyles (not least those reliant on carbon-based growth) that *many* people may value and underlie the workings of contemporary societies, can have detrimental consequences for present and coming generations (people we may meet) as has been made clear by the Intergovernmental Panel on Climate Change (IPCC) (IPCC 2007). Or, to use a case from this book, Rank et al. argue that many chemicals that are used are unacceptable because, for example, they might reduce people's capability to reproduce. Consequently people have *good reason to reject* them and place legal restrictions on their use even if they affect the perceived quality of various shampoos or washing powders.

The idea of reasonable rejection in relation to ethics comes from Harvard philosopher Thomas Scanlon.[4] All moral deliberation involves justification, attempts to convince others, or indeed ourselves, that our actions are right or wrong and one may assume the aim is to find agreement. People may have very different concepts of the good (what is valuable in human life), but by having to justify a moral system to others on grounds that they could not reasonably reject (a desire for reasonable agreement) Scanlonian contractualism gives individuals the power of veto over the imposition of moral systems providing that their rejection is reasonable.

The question then is: What is reasonable rejection? There is no generic answer to this question, such as the avoidance of harm, even if in many cases it will be sufficient (rights for instance may not be directly related to harm). Some cases are quite clear, to use one of Scanlon's examples principles allowing wanton killing can clearly be rejected as those who may be killed have good reason to do so. The issue mentioned above, climate change, is much more complicated as there are many more factors involved. Whilst there is considerable scientific evidence that temperature changes above 2°C may well cause major social and environmental disruptions (IPCC 2007), other considerations also need to be taken into account such as knowledge, realistic opportunities to change energy use and so on. However, given the possibility of such changes, it would be morally wrong not so to do as others would have good reason to reject our inaction.

Scanlon's ideas have also been developed in the field of rights by Brian Barry who in his book *Justice as Impartiality* (1995) points out that, for example, in most usual circumstances it will be difficult to persuade someone that they should have fewer or weaker rights than others. How should we convince someone so that they would freely consent to being denied freedom of speech, or equality before the law, or the right to vote? Interestingly, together with Anand, Sen has endorsed a similar view in relation to sustainable development:[5]

> We have emphasized that sustainability is a matter of distributional equity in a very broad sense, that is, of sharing the capacity for well-being between present people and future people in an acceptable way – that is in a way which neither the present generation nor the future generations can readily *reject*. This is a criterion of justice that has been forcefully used – though not in the context of intergenerational equity by Thomas Scanlon ... and by John Rawls... . (Anand and Sen 2000: 2038) [Emphasis in the original]

---

4   Scanlon and Sen have taught courses together at Harvard.

5   Sen's latest work The Idea of Justice (2009) was published too recently to be incorporated here. Briefly, he rejects Rawlsian contractualism in favour of realisation-based comparison. While generally positive towards Scanlon, Sen's major disagreement concerns the possibility of achieving agreement given diversity of values. However, his objects seem to already have been met by Barry (1995) to whom Sen does not refer.

Anand and Sen's article is the only place where Sen has discussed sustainable development at length. My suggestion here is that if Sen followed that line of thought more consequentially, he would be more cautious in his formulation of development as freedom which gives the impression that the mere addition of freedoms is development. Bringing in the Scanlonian approach provides us with criteria for limiting freedoms and establishing which ones are legitimate.

If development is to be understood as a process of increasing legitimate freedoms, can we categorise some as more basic than others (prima facia the freedom to be nourished is more important than the freedom to buy a Janet Jackson CD)?[6] One of my disgruntlements with the Brundtland Report is that it does not give a clear list of those needs which are important so that we know which needs future generations have that need meeting or, alternatively, the Report could argue why should a list be unwanted. We now turn to this and related questions concerning the capabilities approach within which there is now a substantial discussion concerning lists of capabilities and whether we should have them or not and, if so, which items should or should not be on them (Nussbaum 2002, Alkire 2002, see also *Feminist Economics* 10(3), November 2003). Here, I will confine myself to expounding and criticising Sen's views and proposing and defending mine.

Whilst Sen (1999) has argued that poverty is best conceived as 'basic capability deprivation', he has stoutly refused to give a definitive list of capabilities stating which capabilities are 'basic' and which are to be prioritised. Instead he provides lists, in a similar fashion to the Brundtland Commission, of the kind of thing that he is thinking about, namely deprivations such as those in education, health and not being able to participate in the community. Unlike the Commission, Sen has given several reasons for not providing a canonical list of basic capabilities (Sen 2004, 2005). Firstly, he argues that such a list would not allow for public debate and reasoning to establish it, such a list would be top-down and imposed upon others. Secondly, lists are context specific. To rule out a capability would be automatically to give it a weighting of zero no matter what one investigates. Thirdly, relevant capabilities differ across societies, thus to be able to use the internet might be a relevant capability with respect to participating in a contemporary community, where it was clearly not so a hundred years ago. The list should fit the situation.

There is considerable appeal in this anti-dogmatic approach, however it is particularly problematic in relation to sustainable development as many of the relevant decision makers are not yet alive and consequently do not have voice in contemporary public debates. Furthermore, whilst the argument about public debate is attractive, *engaging in democratic decision making about lists requires that those involved have several capabilities otherwise they could not enter into the debate*. In other words, we cannot avoid having lists of some basic level of well-being.

---

6   But see Sen (2001).

Essential Capabilities: Which and Why

The idea behind the notion of essential capabilities developed here is that they are those that are necessary for deliberative processes enabling the further prioritisation of legitimate freedoms. The term 'essential' is used here to avoid confusion with Sen's use of basic capabilities, a term that would otherwise be more appropriate. Essential capabilities are both individual and collective:

Essential Capabilities

Individual	*Collective*
Bodily health	Reproduction
Agency	Production
Social inclusion	Social capital

We now turn to the *justifications* for designating these essential. Each category is complex and what is included in it may vary from society to society and over time and to this extent the list is necessarily vague. To illustrate, the general category of bodily health involves the absence, as much as possible, of diseases, but it would have been nonsense to stipulate particular diseases before they existed (such as HIV/AIDS) and that degree of specificity cannot be provided. There needs to be some flexibility in lists to account for such instances; however the overall policy goal is clear.

Bodily Health

The inclusion of bodily health is indisputable as the bottom line is survival. International indicators of bodily health usually concentrate on a bio-medical model and include such aspects as under-five mortality rates, life expectancy at birth, the maternity mortality ratio, infectious diseases and nutrition (see for example the Millennium Development Goals). All of these, and more, are clearly important. But there are several inbuilt assumptions which the essential capabilities approach would also highlight. The first is that health starts at birth, this is despite the campaigns informing pregnant women about what they should (e.g. vitamins) or should not eat or drink (e.g. alcohol) during the pregnancy as these things affect a child's life chances later on. Secondly, such indicators avoid the issue of abortion and as part of that gender discrimination in those countries in which boys are preferred to girls and the latter are aborted or, to use another example, the screening for foetal abnormalities is not taken up. Interestingly, a recent chronicle in the Danish Newspaper *Politiken* (17 September 2009) written by a schizophrenic argued against the abortion of those who might develop schizophrenia in later life. This is despite her describing her illness as 'living in hell' as she still has reason to value it. Thus there are some implicit moral judgements that are being made in the usual (re)production of data. Thirdly, bodily health can also be affected by a number of other factors (not just diseases), such as unintentional injuries resulting from road traffic accidents, fires, drowning and so forth, and intentional injuries be they self-inflicted or due to violence such as rape or war. Or again, they may be due to mental disorders such as anorexia nervosa. Fourthly, disabilities are usually completely absent from the picture. Fifthly, as stated earlier, they do not take choice into account.

Agency

As Doyal and Gough have pointed out a body can survive, say on a life support machine, but as such it is not able to do anything and thus mere survival is not enough. There cannot be a leading of valued lives. Agency is usually assumed and not conceptualised within international indicators such as the Millennium Development Goals. It may be defined as the capability to make a difference (Giddens 1984). Giddens, like Bourdieu, has sought to find a middle way between two traditions of understanding agency. At one end of the spectrum, the functionalist/structualist approach has seen agents as being determined by broader social structures, at the other end rational choice theorists and philosophers like Sartre have seen actors as being free and unconstrained by structures. Bourdieu and Giddens see actors as acting within a given context which shapes their actions which in turn shape social structures and systems. The context both enables and restricts action (Giddens 1984, Bourdieu and Waquant 1992). The understanding of context and the weight given to it differs between the two authors. Giddens (1984) is somewhat sketchy in his explanation of structures as rules and resources and much of his more recent academic work has concentrated more on the ways individuals in advanced modernity shape their lives. By contrast the major concepts that Bourdieu employs (field, habitus, forms of capital) emphasise and elucidate the background within which individuals act. Whilst Bourdieu recognises the grounds for charging him with determinism he rejects the acquisitions by stating 'social agents are determined only to the extent they determine themselves' (Bourdieu and Wacquant 1992: 136).

Within the capabilities approach, agency has been seen as central to well-being (Sen 1999). Like Giddens and Bourdieu, though there is no conscious reference, both Sen and Nussbaum have emphasised how preferences, desires, expectations and attitudes are shaped by social backgrounds so that the very capability deprived 'accept their lot'. Their point is more anti-utilitarian rather than sociological: 'The mental metric of pleasure or desire is just too malleable to be a firm guide to deprivation and disadvantage' (Sen 1999: 63). Thus, the capability to make a difference is enabled and limited by social context. One implication of this is that increasing people's agency freedom can require social change and the breaking down of social structures, rules, norms, our upbringing and so forth. In Bourdieu's terms it involves changes in fields, forms of capital and habituses, in some cases this change can be quite radical as in the case of women's subjugation or in relation to the caste system.

Agency as conceptualised here includes the concept of practical reason: 'Being able to form a conception of the good and to engage in critical reflection about the planning of one's life' (Nussbaum 2006: 77). It also involves being able to do things. This highlights the limit to practical reason that mental disorders make and the centrality of mental health to fundamental well-being.

Social Inclusion[7]

As Aristotle pointed out, man is a social animal. Individuals have to live in societies for at least the first four years of their lives in order to survive, those that have lived outside society for a large part of their lives are clearly capability deprived and act more like animals than humans, as was the case with the wild boy of Aveyron (Giddens 1989: 60–3), or suffer from psychosocial disorders (Rosenzweig, Breedlove and Watson 2005: 490). The importance of being able to participate in the community was recognised by Sen in his original Tanner lecture though paradoxically he has done little empirical work in this area (Crabtree 2008). For Rawls (1971: 386), self respect is the most important primary good as it involves an individual's own value as a person, and in the belief in one's own ability to fulfil one's own intentions. That is to say that it is at *the very heart of our being as people*, and therefore it is not surprising that the shaming of others denies them the status of being a human being (Crabtree 2008). It is also central to Honneth's work (see Elling, this volume), and within a Scanlonian framework it is clear that people have good reason to reject not being fully recognised. As to what social inclusion requires depends on the society and is thus necessarily difficult to stipulate, an example being the capability to use Twitter is currently emerging as important for political participation. This category would also include the capability to participate in decision making, rights and non discrimination (Crabtree 2008).

Reproduction and Production

Of the collective capabilities, two appear to be relatively straightforward – societies need to be able to reproduce and produce in order to survive. Clearly it is not necessary for all individuals within a society to produce and/or reproduce and all societies have members such as children who neither reproduce and/or produce. What *is* controversial is the nature of reproduction (what age, how many, for example China's one child policy) and the nature of production (its possible negative effects such as the use of genetically modified organisms or productivism – topics more thoroughly discussed by several authors in this volume). The desirability of growth depends upon *what type of growth* we are discussing. Jelees and Shiva (2006) maintain that the large number of suicides amongst Indian

---

7   One might use Bourdieu's notion of social capital here as, at least on some interpretations, it is something individuals have (or do not) rather than society more generally for which I want to reserve the term.

farmers is due to an unnecessary shift in production methods as Indian farmers have come to rely on costly fertilisers resulting in mounting debts which they cannot repay. One strong criticism I would make of the capabilities approach is that it has concerned itself with opportunities to the extent that it forgets how these opportunities are arrived at, for instance whether nourishment (the capability to be nourished) comes from organic food or not.

Social Capital

At the collective level this refers to social networks and the associated norms of reciprocity and trustworthiness (Putnam 2007). It is metaphorically referred to as the glue that holds society together. Clearly individuals can lack social capital but societies cannot completely (by definition), anarchy and civil war are below the threshold as is exemplified by the Rwandan case. There is a normative question as to whether all social capital is desirable, which the overall tone of many writings on social capital suggests it is (Bourdieu is an exception here). This is not always the case as the strong ties of the Mafia suggest (Putnam 2007). Here again Scanlonian contractualism helps us clarify which societies or parts of them are acceptable or not.

Unlike the Brundtland Report, we now have a definitive list of the essential capabilities which are to be had by and sustained between generations and justifications for their inclusion. Unlike the Brundtland Report, the limitations as to how precise they can be have been indicated. To repeat, the list presented here is *not finite*, there are other legitimate freedoms. Scanlonian contractualism provides the normative base.

To grasp some of the implications of this approach and the differences it makes, we turn to the specific example of mental health and the environment.

Mental health and the Environment

We might assume that the reason why mental health is not on the international agenda is because it is not that big a problem. However, this is not the case.

One of the main documents concerning mental health worldwide is the World Health Organization's 2001 Report *Mental Health: New Understanding, New Hope* (WHO 2001a). Much of the data used in many articles, books and international institutions' reports and working papers comes from either the 2001 Report or the Mental Health Atlases also produced by WHO (WHO 2001b, 2005). The Report characterises mental health as follows:

> Mental health has been defined variously by scholars from different cultures. Concepts of mental health include subjective well-being, perceived self-efficacy, autonomy, competence, intergenerational dependence, and self-actualisation of one's intellectual and emotional potential, among others. From a cross cultural

perspective, it is nearly impossible to define mental health comprehensively. It is however, generally agreed that mental health is broader than lack of mental disorders. (WHO 2001a: 5)

As stated earlier, the idea of understanding mental health in positive terms is endorsed here, but as we are primarily concerned with thresholds and limits to agency we naturally look to the negative aspects. To do so, this chapter follows WHO and the International Classification of Disease 10 (ICD-10) in which mental and behavioural disorders are:

Understood as clinically significant conditions characterized by alterations in thinking, mood (emotions) or behaviour associated with personal distress and/or impaired functioning. Mental and behavioural disorders are not just variations within the range of "normal", but are clearly abnormal or pathological phenomena. (WHO 2001a: 21)

The overall classification of these disorders is as follows:

- organic, including symptomatic, mental disorders, e.g., dementia in Alzheimer's disease, delirium;
- mental and behavioural disorders due to psychoactive substance use, e.g., harmful use of alcohol, opioid dependence syndrome;
- schizophrenia, schizotypal and delusional disorders, e.g., paranoid schizophrenia, delusional disorders, acute and transient psychotic disorders;
- mood [affective] disorders, e.g., bipolar affective disorder, depressive episode;
- neurotic, stress-related and somatoform disorders, e.g., generalised anxiety disorders, obsessive–compulsive disorders;
- behavioural syndromes associated with physiological disturbances and physical factors, e.g., eating disorders, non-organic sleep disorders;
- disorders of adult personality and behaviour, e.g., paranoid personality disorder, transsexualism;
- mental retardation, e.g., mild mental retardation;
- disorders of psychological development, e.g., specific reading disorders, childhood autism;
- behavioural and emotional disorders with onset usually occurring in childhood and adolescence, e.g., hyperkinetic disorders, conduct disorders, tic disorders;
- unspecified mental disorder.

The full specification is given in WHO's diagnostic guidelines (WHO 2007). It is clear that this specification is open to dispute and it would be surprising if the list was never, for the first time in history, revised. The universality of post-traumatic stress disorder (PTSD) is hotly disputed with some claiming that it is culturally relative or a legal concoction which lumps together a range of disorders which would be better understood separately (for a discussion see Van Ommeren et al. 2005). However, given these definitions, we can get some idea of the problem in statistical terms, but it must be said immediately that one of the biggest problems is inadequate data collection (WHO 2001a, WHO 2001b), thus all data are 'best estimates'. These suggest that 25 per cent of the world's population suffer from a mental or behavioural disorder during their lifetime. Even if this estimate is not perfect, it quite clearly indicates the magnitude of the issue and that it should have a central role in policy making and implementation including its connections to sustainable development.

Mental and neurological disorders are estimated to account for 12.3 per cent of disability adjusted life years (DALYs), and 30.8 per cent of all years lived with a disability (WHO 2001a: 25–6). One estimate calculated the total cost to the United States as being 2.5 per cent of gross national product (WHO 2001b: 27). The evidence suggests that income poor people have double the number of disorders than non income poor, the causes are related to poor nutrition, education, unemployment, homelessness, disease, minority status, migration, conflicts, violence and being cut off in rural areas. Furthermore, mental disorders also cause poverty as families often have to bare the burden of the costs. In low income countries 42.9 per cent of costs are out of pocket expenditures (WHO 2005). Furthermore, jobs are lost and there is reduced productivity from the care-giver who has less time to work. Crime rates can also increase. The greatest 'cost' is, of course, to those who have the disorders and their families. In addition to the disorder they often face stigmatisation and discrimination, and find their human rights reduced – liberty is curtailed through detention.

Despite the clear importance of the problem, the links to sustainable development have not been discussed (there is, for example, scant mention of mental health in the health section of the UNEP's Millennium Ecosystem Assessment). What then are the connections between mental health and the environment? Figure 8.1 provides an analytical framework for the kinds of interactions we can find. It builds on WHO's conceptualisation of the various factors that affect mental and behavioural disorders – biological, psychological and social – to which an outer ring is added viz the (rest of the) environment.

The aim in this chapter is not to go through every mental disorder and specify the interrelations; this would be a massive task. For example, there are 12 known viruses which affect development disabilities alone (Institute of Medicine 2001: 129–31). Rather the aim is to exemplify the interrelationships and indicate the kind of problems which can arise when trying to understand them. As the diagram suggests, environmental factors belong to two categories: (1) the body; and (2) the rest of the environment. These link into other factors, for example according to one study, following hurricane Katrina there was a *decrease* in suicide ideation (Kessler et al. 2007). The suggestion was that this was due to the increased social capital (closer ties) which had developed. (This finding contrasts with a statement by Dr Jeffrey Rouse, the deputy New Orleans coroner dealing with psychiatric cases, who

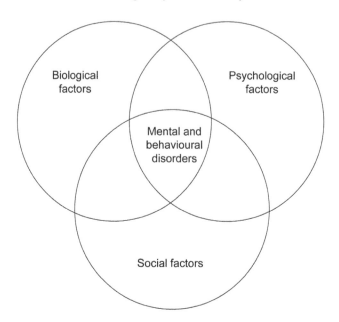

Figure 8.1   Conceptualising the relationship between mental disorders and the environment

said that 'the suicide rate in the city was less than nine a year per 100,000 residents before the storm and increased to an annualized rate of more than 26 per 100,000 in the four months afterward, to the end of 2005' (*New York Times*, 21 June 2006).

Biological Factors

As stated earlier, humans are social *animals*, that is, we are part of the 'natural' environment. To start with the body, perhaps the most obvious link is the genetic one. Schizophrenia is now known to have a genetic component, although this is not the only factor in the development of the disorder. The risk of developing schizophrenia is 48 per cent if one is a monozygotic twin and one's twin suffers from it, the risk in relation to the general population is 1 per cent (Rosenzweig, Breedlove and Watson 2005: 490). But not all genetic disorders are inherited, for example Down Syndrome (mongolism) is a result of an extra chromosome due to the faulty separation of two chromosomes in the egg cell before ovulation. Unlike schizophrenia, there is no known risk of the disorder being passed on to the next generation (Hilgard, Atkinsen and Atkinsen 1975: 421). Perinatal conditions can also have consequences, for example premature births can lead to cerebral palsy and cognitive disabilities (Institute of Medicine 2001: 132). There is evidence to suggest that hormonal changes may contribute to post-natal depression amongst women, though it is clear that other factors also play a role (Institute of Medicine

2001: 297). Recent advances in neuroscience have underlined the connections between the brain's physiology and mental disorders. For example, Parkinson's disease is linked to the lack of dopamine in the brain (WHO 2001a).

The Rest of the Natural Environment

As White and Heerwagen (1998, 175) state 'Given our species long history as subsistence hunters, gatherers and farmers, the natural environment must have helped shaped our cognitive and emotional apparatus'. Several factors fall under the general rubric of 'the natural environment' examples being animals, nutrition, infections and natural disasters. There are a number of recognised phobias relating to animals (such as spiders and snakes) and the natural world, especially if we include height and space. Nutritional intake can have serious consequences for mental health. For example, it is estimated that in India alone 70 million people have an iodine deficiency disorder and 2,000 million are at risk (Prahalad 2001). The deficiency can lead to cognitive, motor and hearing disabilities and in some of the worst cases cretinism (Institute of Medicine 2001: 126). Infections (parasitic, bacterial and viral) are many, but some of the most important are malaria, which exists in approximately 90 countries and affects several hundred million people, which can cause cerebral malaria (leading to convulsions, drowsiness and comas) and related depression. HIV can cause the central nervous system to be impaired and cognitive and movement disabilities. It can also lead to seizures – approximately 60 per cent of those with HIV in Africa suffer from epilepsy (Institute of Medicine 2001: 129, 189).

Although work is now being done on the mental health effects of climate change not least in relation to extreme weather events, the IPCC (2007) in its review of the health outcomes of climate change has pointed out that there are relatively few studies of the psychological consequences, most of the studies that have been undertaken have been done so in high or middle income countries (Ahern et al. 2005, Ginexi et al. 2000). In fact, one extensive literature review undertaken recently could only find nine studies relating flooding to mental health in developing countries. Reviews of the literature that do exist reveal that common mental disorders resulting from extreme weather events have been found in a number of countries, and different cultural and social contexts (IPCC 2007, Norris et al. 2006).

And the Other Way

The environment not only affects mental health, mental health also affects the environment. Poor mental health can affect the rest of the body, indeed WHO goes so far as to claim that 'mental health is fundamentally linked to physical health outcomes' (WHO 2001a: 10). This is fairly clear from addictive disorders like smoking, drinking and drug abuse (and subsequent HIV/AIDS problems), but mental disorders have also been linked to other outcomes. For example, depression has been found to affect heart disease (WHO 2001a). The links in relation to the rest of the environment are less clear. Most obviously, debilitating disorders such as Alzheimer's disease mean that suffers cannot play an active role in relation to the environment. The demand for various addictive drugs can affect land use patterns, and consequently, ecosystems. Irrigation might be required, for example, for the growth of opium poppies, and manufacturing and transportation processes may also have negative effects (see Ghotbi and Tuskatani 2007). Clearly it is also an area for further research.

Such cases draw our attention to the variety of actions that need to be undertaken if we are to live sustainably. The cases above are by no means exhaustive, but they do show the kinds of relationships that can exist between the environment and mental disorders.

However, they only trace the relationship between the environment and mental health functionings.[8] The view taken here is that development is the process of increasing people's legitimate freedoms (capabilities). Thus, it is important when looking at the model that the social aspects include, unlike WHO's model, social factors that affect those with mental disorders such as treatment opportunities, stigma, rights and so forth which influence individuals' capabilities to live the lives they value.

Conclusion

This chapter has argued for the replacement of the basic needs approach in relation to sustainable development. It has argued that future generations should have the generalised capacity to produce well-being here identified in terms of essential capabilities. Development does not end with the attainment of essential capabilities. It also involves an increase in legitimate freedoms, a concept which is founded on Sen's capabilities approach and a Scanlonian approach to ethics and rights. Thinking in these terms highlights issues not normally foregrounded in the sustainable development debate despite their significance. This is the case with mental health which is important to the essential capability of agency. After outlining the significance of the problem, its relationship with the environment has been sketched out as the basis for further thinking in this area. Work we owe to present and future generations.

References

Ahern, M., Kovats, R.S., Wilkinson, P., Few, R. and Matthies, F. 2005. Global Health Impacts of Floods. *Epidemiological Review*, 27, 36–46.
Alkire, S. 2002. *Valuing Freedoms*. Oxford: Oxford University Press.
Anand, S. and Sen, A. 2000. Human Development and Economic Sustainability. *World Development*, 28(12), 2029–49.

---

8   The word 'functioning' is particularly inept in this context because of its mechanical connotations, however, no such connotations are intended (Sen pers.com).

Barry, B. 1991. *Liberty and Justice.* Oxford: Clarendon Press.

Barry, B. 1995. *Justice as Impartiality.* Oxford: Clarendon Press.

Bourdieu, P. and Wacquant, J.D. 1992. *An Invitation to Reflexive Sociology.* Oxford: Polity Press.

Crabtree, A. 2008. The Centrality of Basic Social Capabilities, in *Agir et réagir ensemble: Capabilités collectives, Contribution du Réseau IMPACT*, edited by J.-L. Dubois, A-S. Brouillet, P. Bakhshi and C. Duray-Sounron. Paris: L'Harmatton.

Department for Environment, Food and Rural Affairs (DEFRA). 2007. *Sustainable Development Indicators in Your Pocket.* London: DEFRA Publications.

Doyal, L. and Gough, I. 1991. *A Theory of Human Need.* London: Macmillan.

Ghotbi, N. and Tuskatani, T. 2007. *The Economy of Opium and Heroin Production in Afghanistan and its Impact on HIV Epidemiology in Central Asia, Discussion Paper No. 635* [Online: Kier Discussion Paper Series, Kyoto Institute of Economic Research]. Available at: http//:www.kier.kyoto-u.ac.jp/index.html [accessed: 15 August 2009].

Giddens, A. 1984. *The Constitution of Society.* Oxford: Polity Press.

Giddens, A. 1989. *Sociology.* London: Polity Press.

Ginexi, E.M., Weihs, K. and Simmens, S.J. 2000. Natural Disasters and Depression: A Prospective Investigation of Reactions to the 1993 Midwest Floods. *American Journal of Community Psychology*, 28, 495–518.

Hilgard, E.R., Atkinson, R.C. and Atkinson, R.L. 1975. *Introduction to Psychology.* 6th Edition. New York: Harcourt Brace Jovanovich.

Institute Of Medicine. 2001. *Neurological, Psychiatric, and Developmental Disorders.* Washington, DC: National Academy Press.

Intergovernmental Panel on Climate Change (IPCC). 2007. *Climate Change 2007.* Cambridge: Cambridge University Press.

Jalees, K. and Shiva, V. 2006. *Seeds of Suicide.* New Delhi: Navdanya.

Kessler, R., Galea, S., Jones, R.T. and Parker, H.A. 2007. *Mental Illness and Suicidality After Hurricane Katrina.* WHO Bulletin.

McKensie, K., Whitley, R. and Weich, S. 2002. Social Capital and Mental Health. *British Journal of Psychiatry*, 181, 280–83.

Norris, F.H., Galea, S., Friedman, M.J. and Watson, P.J. 2006. *Methods for Disaster Mental Health Research.* London: The Guilford Press.

Nussbaum, M.C. 2001. *Women and Human Development.* Cambridge: Cambridge University Press.

Nussbaum, M.C. 2003. Capabilities as Fundamental Entitlements: Sen and Social Justice. *Feminist Economics*, 9(2–3), 33–59.

Nussbaum, M.C. 2006. *Frontiers of Justice.* Cambridge: Harvard University Press.

van Ommeren, M., Saxena, S. and Saraceno, B. 2005. Mental and Social Health During and After Acute Emergencies: Emerging Consensus? *Bulletin of the World Health Organization*, 83(1), 71–6.

Prahalad, C.K. 2006. *The Fortune at the Bottom of the Pyramid.* Upper Saddle River, NJ: Wharton School Publishing.

Putnam, R. 2007. E Pluribus Unum: Diversity and Community in the Twenty-first Century. *Scandinavian Political Studies*, 30(2).

Rawls, J. 1999. *A Theory of Justice*. Oxford: Oxford University Press.

Reader, S. 2006. Does a Basic Needs Approach Need Capabilities? *The Journal of Political Philosophy*, 13(3), 337–50.

Rosenzweig, M.R., Breedlove, S.M. and Watson, N.V. 2005. *Biological Psychology*. Sunderland: Sinauer Associates.

Roybens, I. 2005. The Capability Approach – A Theoretical Survey. *Journal of Human Development*, 6(1), 93–117.

Scanlon, T.M. 2000. *What We Owe Each Other*. Cambridge: Harvard University Press.

Sen, A. 1980. Equality of What?, in *The Tanner Lecture on Human Values*, edited by S. McMurrin. Cambridge: Cambridge University Press.

Sen, A. 1999. *Development as Freedom*. Oxford: Oxford University Press.

Sen, A. 2001. What's the Use of Music? The Role of the Music Industry in Africa. [Online]. Available at: http://siteresources.worldbank.org/INTCEERD/Resources/CWI_music induy_in_Africa_synopsis.pdf [accessed: 15 August 2000].

Sen, A. 2003. Capabilities, Lists, and Public Reason: Continuing the Conversation. *Feminist Economics*, 9(2–3), 319–32.

Sen, A. 2005. Human Rights and Capabilities. *Journal of Human Development*, 6(2), 151–66.

Sen, A. 2009. *The Idea of Justice*. London: Penguin.

UNEP. 2003. *Ecosystem and Human Well-being: Millennium Ecosystem Assessment*. London: Island Press.

White, R. and Heerwagen, J. 1998. Nature and Mental Health: Biophelia and Biophobia, in *The Environment and Mental Health: A Guide for Clinicians*, edited by A. Lundberg. Mahwah, NJ: Lawrence Erlbaum Associates.

Wiggins, D. 1987. *Needs, Values, Truth*. Oxford: Blackwell.

World Commission on Environment and Development. 1987. *Our Common Future*. Oxford: Oxford University Press.

World Health Organization. 2001a. *Mental Health: New Understanding, New Hope*. Geneva: World Health Organization.

World Health Organization. 2001b. *Mental Heath Atlas 2001*. Geneva: World Health Organization.

World Health Organisation. 2005. *Bulletin of the World Health Organization*, 83(1).

World Health Organisation. 2007. *International Statistical Classsification of Diseases and Related Health Problems. 10th Revision Version for 2007*. [Online]. Available at: http://aps.who.int/classifications/apps/icd10online [accessed: 22 February 2010].

Chapter 9
Economic Analysis of Sustainable Development

Anders Chr. Hansen

Introduction

From an economic point of view, sustainable development is about balances between present and future, and between environmental qualities and economic activities. The standard model identifies the optimal balance between environmental qualities and consumption of produced goods, where these balances are based on the mainstream economic approach to assessing the balance between environmental qualities and the production of goods describes the trade-off with a standard model such as the one shown in Figure 9.1. The intention is to introduce to the most fundamental types components of the balance, but it is often used beyond this end. It also serves directly as a guide for empirical research by students and even sometimes by graduated specialists. In the words of Schumpeter, it is used as a *preanalytical vision.* It specifies the kinds and forms of relations that we can expect to find and therefore the kind of relations to look for. Using the simple standard model for that purpose may lead to overlooking important economic potentials and to neglect important properties of the problem known in other academic fields, such as science, technology, philosophy and political science.

This chapter discusses how the standard models can be adapted to better serve as a preanalytical vision in analysis of sustainability and environmental-economic choice problems that are interdisciplinary by nature. It is aimed at an interdisciplinary audience with varying theoretical backgrounds. Therefore, the models are only shown graphically and they are accompanied with brief explanations of fundamental economic concepts.

Sustainable Development and Economic Analysis

Sustainable Development and the Environment and Growth Debate

The Brundtland Commission (World Commission on Environment and Development (WCED) 1987) and the Rio Declaration (United Nations Conference on Environment and Development (UNCED) 1992) coined the concept of *sustainable development* as a global endeavour, including economic, social as well as environmental progress. Moreover, the fact that any generation, knowingly or unknowingly, dictates the economic, social and, in particular, environmental conditions of its descendants makes inter-generational ethics fundamental to the concept of sustainable development. According to WCED (1987), the core principle of sustainable development is the 'development that meets the needs of the present without compromising the ability of future generations to meet their own needs'.

This idea of *sustainable development* can be seen as a synthesis of the Great Debate in the 1970s of whether continuing economic growth is physically possible on a finite planet. Probably the most debated contribution was the Limits to Growth Report (Meadows et al. 1972) warning that with the prevailing *energy use, pollution and population patterns* of the industrialised economies, the planet would not be able to sustain a similar industrialisation and economic growth of the Third World – let alone continuation of growth in the industrialised world. Today, these concerns are mainstream policy and the common goal of sustainable development was a way to understand, express and react to them across cultures, political divides and generations.

Zero economic growth was also suggested as a solution in the Great Debate and sustainable development is a denial of that solution. First, economic growth is necessary to bring the poor billions of the world population out of poverty. Second, even if economic growth in the rich countries was stopped in, say, year 2000, their economies would still be unsustainable and not a single environmental problem would be solved. Sustainable development is about conversion and transition to economic activities that are in long-term balance with the environment and these activities must necessarily be growing.

The economic understanding of growth is slightly different than the understanding of growth in natural science. Whereas science defines growth as accumulation and increased density of matter and energy, growth has a different meaning in economics. Economic growth refers, in principle, to the *value added* to the energy and matter used in economic activities not to energy and matter per se. From an economic point of view, it is clearly possible to create more value while using less materials and energy. A car that uses 5 litres of petrol per 100 km is more valuable than a corresponding car using 10 litres per 100 km although it contains less steel. It does contain more engineering and probably more excellence in manufacturing and thus the production as well as the consumption of it uses more economic resources, but less steel and fuel while delivering the same mobility to the car user.

Sustainable development involves shifts from low value/high environmental pressure to high value/low environmental pressure production and consumption. In Europe, this has been called *delinking* or *decoupling* of throughput growth from economic growth (e.g., Council of the European Union 2006).

Delinking represents an enormous challenge. Energy-efficiency and eco-efficiency[1] must grow at least at the same rate as the economy, just to keep total

---

1   Measured as value added/energy demand and value added/emissions.

energy use and environmental pressure from rising. To reduce energy use and environmental pressure to sustainable levels, even higher rates of energy- and eco-efficiency growth are necessary. Such rates of progress are possible, but demanding. The average annual growth rate of greenhouse gas emission efficiency in Denmark was 5.8 per cent in 1995–2000 (Hansen 2002a). This could, however, only come about due to an intensive use of green taxes, regulation, support for cleaner alternatives and infrastructure investments. Generally, delinking cannot be expected to take place within the same economic institutions and political priorities that led to the close link between throughput growth and economic growth in the past. Institutional change is key to sustainable development.

Economics and Sustainable Development

The insights needed for institutional change come from science and humanities as well as technological knowledge and economics and must be taken into consideration across academic disciplines. It is multidimensional and interdisciplinary. Nevertheless, transition to sustainable development involves questions about economic sustainability, optimality and efficiency, distribution and economic institutions. Economics as an academic discipline has evolved to provide answers to questions of exactly these kinds. This chapter focuses mainly on the first two classes of questions.

Red figures on the account balances are indicators of economic *unsustainability*, at least if they persist. At the macro level, national accounts and growth models serve as tools for addressing this question. In the long run, green and dirty practices alike have to be economically balanced to be viable and economic growth has to regenerate its resource base to be sustainable. The kind of balances required for economic sustainability resembles those required for environmental sustainability.

Another class of questions addressed by economics concern the *optimality and efficiency* of alternative sets of inputs and outputs of scarce goods and their relative prices or costs. All households, firms, authorities and governments are concerned with how to get the most out of their resources. Cost-Effectiveness-Analysis and Cost-Benefit-Analysis are the archetypes of economic tools for economic appraisal and properly carried out they are indispensable components of multi-criteria analysis.

Sustainable Use of Nature

Nature is a resource base for the economy directly as environmental support for human life, health and other factors of well-being and indirectly as a source of input to production, e.g., fisheries, forestry and mineral extraction.

In fisheries and forestry, the annual amount of fish that can be caught or timber that can be logged on a permanent basis cannot exceed the annual rate of regeneration of the fish or timber stock. Thus, the sustainable harvest depends on the stock of the resource and its rate of regeneration. To identify which of the sustainable sets of stock, regeneration and harvest are optimal from an economic point of view, the value of the harvest must be compared with the cost of harvesting.

Mineral resources are exhaustible or non-renewable with no regeneration. In this case, it is neither possible to sustain a particular stock of the resource nor of the extracted amount indefinitely. In principle, the resource belongs to all generations, but it has to be extracted by only some of them. This problem was discussed in depth in the economic literature during the Great Debate (Dasgupta and Heal 1974, Solow 1974, Stiglitz 1974). The answer to the question was to invest the proceeds from the resource in man-made capital – later on spelled out in the 'Hartwick rule' (Hartwick 1977). In this way, it is possible to sustain a certain level of consumption, even when the resource stock is fully exhausted.

The fundamental assumption in this line of thought is, as noted above, that natural resources are only *natural capital* which can be frictionless and substituted by *man-made capital* (buildings, infrastructures, machines, transport equipment). Non-renewable resource extraction is like selling pieces of one's property and it can be balanced by investing correspondingly in man-made capital leaving the *total capital* stock (man-made + natural capital) unchanged. This sustainability criterion was extended to include the destruction of environmental qualities in subsequent papers (Hartwick 1994). Later on, it was labelled 'weak sustainability' (see below). The assumption of perfect and frictionless substitution of environmental values by economic values was already by then frequently used in economic literature. Often with reference to a notion that this was the only way environmental values could be taken into account at all.

The substitutability assumption rests on the notion that the consumption of produced goods is the ultimate purpose of other economic activities, production and investment. These activities have no other justification than being necessary for consumption. The purpose of consumption itself is human well-being. Natural resources and environmental qualities enter this hierarchy of values either through their importance for the ability of the economy to produce goods or through their direct importance for human well-being. In their capacity of inputs to production, natural resources are not more or less valuable than the inputs they replace or can be replaced by.

The inference from this view is that it is the human well-being that should be sustained, not production and investment and neither natural resources and environmental qualities. They are all just means to directly or indirectly provide human well-being. Furthermore, if we can measure well-being in monetary units, we can find the cash equivalent of every contribution to this well-being producing system. The contribution from natural resources to productive capacity can be replaced with contributions from man-made capital.

The 1974 papers were primarily concerned with the question of whether a given level of consumption can be sustained when the resource input eventually declines as the non-renewable resources are exhausted. They concluded that it depends on whether they are *substitutable* and *indispensable* or only one of the two. If the resource is indispensable and not fully substitutable, production will cease at some time. If it is dispensable, the resource problem per se is not a theoretical problem, but a practical problem of transition to the alternatives. If it is indispensable and substitutable, sustainable growth or a sustainable level of consumption requires an ever increasing amount of man-made capital to compensate for a permanent decline of resource inputs.

The assumption of infinite substitutability was subject to a heated debate involving proponents declaring an 'age of substitutability' (Goeller and Weinberg 1976) and opponents criticising the assumption for being inconsistent with the physics of transformation of energy and matter (Georgescu-Roegen 1971, Daly 1992, 1997). Much of the debate is centred on the scarcity of fossil fuel resources, in particular oil and natural gas. The critics point to the fact that the substitutability of other scarce minerals to a very high degree depends on the availability of inexpensive energy. As long as finite and shrinking fossil fuel reserves remain the primary energy basis of our economies, the infinite substitutability assumption seems less realistic.

Recent advances in energy efficiency and renewable energy technology, however, give sufficient reason to believe that fossil fuels are indeed dispensable and that European countries are developing strategies for transition to a *post-carbon* or *non-fossil* future. This alternative was present already in the Great Debate, but it took a generation for the public opinion in Europe to move to the position that it is a *realistic* and socially preferable alternative.

Thus, the *physical* dimension of the sustainability problem is not how to continue production and growth as supplies from mineral resources become increasingly constrained. Rather, it is a matter of transition to sustainable alternatives. The *financial* dimension, however, is important as natural resources also represent *national wealth* that future generations from an ethical point of view are as entitled to as the present generation. In economic models, the financial problem is much simpler to solve.

Criteria for sustainable consumption can be derived from national accounts data. According to the weak sustainability criterion, sustainable consumption cannot exceed the net national product adjusted for resource consumption. Extracted reserves represent a one-time sell-off of assets. Not a sustainable source of income. Alternatively, the resource extracting sector can be separated from the other sectors as the 'golden rule principle' suggested by among others (Hansen 2001). The net product of the resource sector is then cumulated in a resource fund. Sustainable consumption will be the net product from the other sectors plus the returns from the resource fund.

The latter approach is very close to the fiscal rule in force in Norway from 2002. It states that the structural deficit of the government budget except its oil and gas revenues can be financed by up to four per cent of the value of the Petroleum Fund (Finansdepartementet 2001). The capital of the fund can potentially be channelled into investments in transition, but also into any other investment opportunity worldwide.

Other countries have had more difficulties disciplining their consumption of oil wealth and investing the proceeds in future. This was, for instance, the verdict of the Russian President on the state of the Russian economy – actually a flaw going back to Peter the Great (Medvedev 2009). Resource rich countries that are already industrialised often suffer from symptoms of the 'Dutch disease' (at least in passing), but have no persistent problems of keeping their total capital stock constant while depleting their reserves. However, the rate of investment in many resource rich low income countries is often persistently too low to sustain even the low level of consumption in these countries (Hansen 2002b).

Strong Sustainability

As noted above, it now seems possible for other energy sources and technologies to replace the specific qualities of fossil fuels, such as the excellent suitability for transport purposes of the high energy density of petrol and diesel in combination with a combustion engine. With sufficiently abundant sources of non-fossil energy, the substitutability of other mineral resources is also a much more realistic assumption. Environmental qualities, however, are different. Environmental losses such as the extinction of ecosystems and species, cumulated hazardous chemicals in nature or greenhouse gasses in the atmosphere are often irreversible and go beyond the bounds of acceptable living conditions for man and nature. Such losses are losses of goods that unlike natural resource inputs are 'consumed' directly. Thus, the idea of compensating the use of non-renewable resources by higher investments was developed further by distinguishing between *weak* and *strong* sustainability (Pearce and Turner 1991, Neumayer 2003, Atkinson, Dietz and Neumayer 2007).

The strong sustainability criterion would include preservation of a certain minimum level of *critical natural capital*. Critical capital is natural capital stocks that cannot be replaced by man-made capital because it does more than yielding a return. Such stocks are critical for sustaining life and health of whole populations or even humanity as well as some of the ethical and aesthethic values that cultures are made of. A good ecological quality is necessary to maintain biodiversity and biodiversity itself sustains good ecological quality. Climate change, if unchecked, could change fundamentally, which parts of the planet are habitable. Unique pieces of nature such as the Grand Canyon are not substitutable because they are unique, but whether they are indispensable for our and humanity's well-being in the future is an ethical and aesthetic rather than an economic question.

Thus the 'critical' part of the concept strips the 'capital' part of an important property: assets that have market value, that is, can be exchanged by equivalent assets. Nevertheless, even these critical natural resources are stocks that yield a flow of services and are in this respect comparable to capital.

Two key properties of critical capital are its *importance* and the *threat* to its existence (De Groot et al. 2003). It takes a multitude of socio-economic and ecological criteria to describe its importance and the threat to its existence is a matter of future sustainability as well as more immediate threats.

But what does it mean to maintain such a stock of critical capital? An operational procedure could be to identify the resource qualities and quantities necessary for maintaining some pre-defined sustainability standards (Ekins 2003, Ekins, Folke, and De Groot 2003, Ekins and Simon 2003, Ekins et al. 2003).

However, even if we follow these suggestions, it remains a political question to define the *sustainability standards* and consequently the maximum levels of environmental pressure compatible with these. Which are the environmental qualities we really want to sustain and which would we accept to abandon in return for economic benefits? Or how much consumption opportunities are we prepared to abandon in return for restoration of natural amenities? Thus, sustainability standards are outputs of the political decision-making process as much as inputs from science.

For the global climate, the sustainability standard could be to avoid runaway climate change implying a maximum temperature increase of 2°C.

Weak and strong sustainability concepts are not totally mutually exclusive in actual long-term strategies. Norway, for instance, converts the oil and gas wealth to economic assets in a Petroleum Fund, the returns from which, in principle, can lift the government budget indefinitely and the investments of which can finance much of the investments needed for transition.

In this perspective, the interesting economic problem becomes how to choose between wide ranges of scenarios that are possible within these constraints. Which are socially preferable?

Optimality or Sustainability?

The Environmental-economic Standard Model

The economic approach to this question takes departure in the standard model for environmental-economic choice. This model only represents the most fundamental balances involved in ecological-economic choice. Nevertheless, it is often used as a preanalytic vision guiding empirical analysis. In the following, the model will be examined and adapted to analysis of ecological-economic choices in the transitions to sustainability.

The standard model takes its departure in the general model of supply and demand in the market. The options for 'supply' of pollution abatement or, more generally, reduction of environmental pressure can be arranged according to the marginal cost of abating an extra unit. It is assumed that low cost options will be applied before more expensive options. Thus, if environmental pressure is already significantly reduced, the marginal costs of further abatement will be high, whereas if little has been done yet, they will be low.

Without government intervention, there is no demand for reducing environmental pressure, but if there was, it could be ordered in a similar way. Society can be assumed to be willing to pay much for a unit of abatement, if very little is abated and the environmental quality consequently is very poor. If the environmental pressure already has been reduced and the environmental quality is good, society will pay less for an additional unit of abatement.

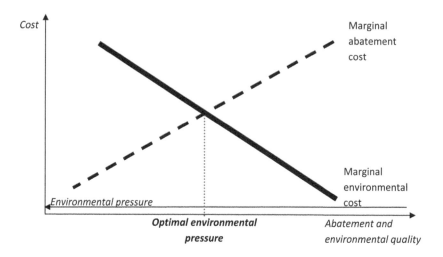

Figure 9.1   Optimal environmental pressure in the standard model

The model is shown in Figure 9.1. The environmental quality and abatement effort is represented by the horizontal axis whereas the environmental pressure runs in the opposite direction. Thus, a high level of environmental pressure equals a low level of environmental quality and a low level of abatement.

It follows from the model that the optimal environmental pressure will be exactly where the marginal abatement cost equals the marginal willingness to pay for an extra unit of abatement. To the right of this point the cost of additional abatement will exceed the gains from it. Thus, it would not be rational to do it. To the left, the opposite would be true.

The simplicity of the model shown in Figure 9.1 is charming, but it gives rise to a series of controversies between economists and natural scientists, sociologists, etc. How does *optimal* relate to *sustainable* environmental pressure? The model shows the socially desirable level of environmental pressure, but not the transition to get there. It implies that environmental values such as an environment free of threats to human health and life and as wonders of nature can be measured in euros or other currencies. These and several other questions will be addressed in the following by elaborating on the causal relations of the model one by one and, finally, putting it all together.

The Marginal Abatement Cost Function

The marginal abatement cost function links the cost of additional abatement activities with the abatement efforts already undertaken. Abatement can be any action that helps reduce environmental pressure. It could be end-of-pipe solutions such as installing filters or changes in technology used such as raw materials with less harmful waste products. Or, just to focus more on reducing unnecessary wastes and use of environmentally damaging inputs. The model is primarily developed for analysing local pollution problems, but it is also used widely to analyse transitions of sectors and the economy to sustainable environmental pressure. In that case, concepts like *transition costs* or *environmental pressure reduction costs* are often used interchangeably with *abatement costs*. The underlying axiom is that production at lower levels of environmental pressure must be less than production at higher levels. This is because capital and labour are allocated to abatement rather than to production.

Consequently, society faces a trade-off between opportunities for consumption of produced goods and environmental quality. Inspired by the debate on mitigation costs in the 1990s (Hourcade and Robinson 1996) such a trade-off is shown in Figure 9.2.

The *production possibility frontier* describes the combinations of consumption of produced goods and environmental quality that are *theoretically* possible. At any point *on* the frontier we can only get a higher environmental quality by reducing our production of goods. Less nitrogen pollution from pig farming would, for instance, only be possible by reducing the number of pigs produced. Less emissions of CO_2 from electricity and heat generation would only be possible by cutting of electricity for some hours and by lowering the room temperature in winter.

Every point inside the production possibility frontier is also possible, but the points inside the curve – such as that represented by the star – are technically inefficient. In this area, it is possible to consume more produced goods without reducing environmental quality and vice versa. Mass unemployment is a macro-economic sign of idle resources, which, in principle, could be used to produce

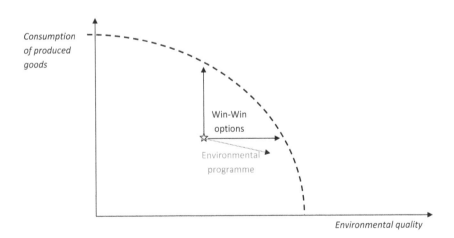

Figure 9.2   Theoretical and actual trade-off between produced and environmental goods

more of both. On the micro scale, numerous sources of environmental pressure are not configured and operated according to their maximum economic performance. Some of the reasons for this are listed below.

The technologies and institutions behind the production possibility frontier are given. When we use it to compare results before and after an environmental programme, however, we introduce a dynamic dimension because transitions always take time.

In macro-economic analysis of transition scenarios, it is often assumed that the economy operates at the production possibility frontier. However, studies of energy and environmental efficiency in production and households have repeatedly found that many households and firms operate within the curve (Jaffe and Stavins 1994b, a, Porter and Linde 1995, Intergovernmental Panel on Climate Change (IPCC) 1996, 2001). They simply use more energy, more raw materials, etc. than necessary to produce their services and output. They can lower environmental pressure *and* cut down on expenditure at the same time, but they hesitate doing it. This phenomenon is referred to as the *efficiency gap* or the *efficiency paradox*.

Why would firms and households use more energy than they have to? Some explanations point to the fact that adopting new and better technologies necessarily takes time, whereas others point to various reasons for households and firms to be reluctant to use them. The reluctance can be due to lack of sufficient technical and economic assessment capacity, low confidence in professional advice, new technical solutions or financial arrangements, procrastinating behaviour or short-sightedness. Government failures such as a practice of granting perverse subsidies may also work counter to the use of economically and ecologically efficient solutions.

Accounting for such deficiencies can fill some of the gap and there is also some scepticism towards the notion of win-win options as such. However, even if this position is accepted, we must expect the current (2009) capital stock to be optimised on false premises. Until 2008, the general expectation by the International Energy Agency, governments and markets alike was that the future oil price would be $30–60 per barrel in the 2010–2030 period. In 2008, the International Energy Agency (IEA) made a detailed account of all oil reserves in the world and found that they are smaller than previously thought and that supply will be correspondingly lower and the oil price more likely $100–150 per barrel (IEA 2008). This is important for use and production of energy because the price of natural gas, electricity and to some extent coal follows the oil price. These new expectations are increasingly shared by governments and markets. All investments in buildings that are to be heated or cooled, machines, engines and transport equipment that would be optimal at an oil price of $30–60 per barrel will be very sub-optimal at an oil price of $100–150 per barrel. The fact is that almost all of our physical capital stock is invested earlier and under these false premises.

In European countries, government policies subsidising agriculture and food production, shipbuilding and shipping, air transport, coal mining, etc. represent other sources of inefficient resource use. They give incentives to tie up labour, capital and land in occupations with very low value creation and very high environmental pressure.

The standard marginal abatement cost curve is easily derived from the production possibility frontier. If we start from the left with very low environmental quality and very little abatement, it is possible to increase the environmental quality by giving up a little of the produced goods. As more pollution is abated and thus a higher environmental quality is achieved, still more production must be given up. The result is an abatement cost curve that increases with abatement. But still assuming that the economy produces the maximum output possible at a given environmental quality and that every possibility of protecting the environment at a given production level is fully exploited.

As noted above, the marginal abatement cost curve can be derived as the tangent slope along the production possibility curve. This is, however, the theoretical marginal abatement cost curve. If inefficient solutions exist, a given programme for improving the environmental quality, say, of the aquatic environment or the atmosphere, will be able to move the actual resource allocation closer to the production possibility curve while improving the environmental quality. If the dotted arrow represents the transition of the economy following such a programme, the slope of the arrow represents the marginal abatement costs and they will be much lower than the theoretical marginal abatement costs. Analysing such transitions to higher value creation and lower environmental impact is maybe the most challenging task to environmental economic analysis.

In Figure 9.3, the distinction between the actual and the theoretical marginal cost curve is shown although really nothing is known a priori about the distance between them, only that we must expect to find actual marginal abatement costs that are lower than the theoretical costs.

If we accept the notion that not all resources of the economy are allocated efficiently, the actual abatement costs will be lower than the theoretical abatement costs corresponding to a given level of abatement and environmental quality. They can even be negative as shown in Figure 9.3.

The cost of reducing environmental pressure can show significant *economies of scale* at the firm level as well as on the industry or sector level. Often fully new plants can offer lower costs and lower environmental pressure at the same time whereas smaller abatement efforts at existing older plants are costly. Infrastructure for electrical vehicles will be very expensive per vehicle in the beginning because it takes some time to build up a car fleet sufficient for an appropriate capacity use of the infrastructure. A large car fleet can benefit from a larger infrastructure network than a small one. Such effects are illustrated by the downward bending of the right part of the marginal abatement cost curve in Figure 9.3.

Abatement costs are also subject to *dynamic* economies of scale. This is important if the analysis is about the reduction of future environmental pressure. New technologies become less costly as firms learn how to optimise their use and production. This means to some extent that the costs of abatement in the future depend on the amount of abatement at the present. Numerous studies have

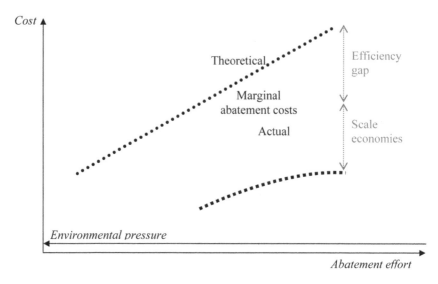

Figure 9.3   Theoretical and actual marginal abatement cost function

documented learning effects for firms and experience curves for industries, but they are difficult to predict. However, it is typically more likely that future marginal abatement cost choices will be positioned lower in the diagram than to assume that it remains in the initial position (e.g., OECD and IEA 2000).

Thus, when analysing transitions to environmentally sustainable economic activities, the most rewarding contribution of economics is not to demonstrate how the economy can run back and forth on the production possibility curve. Rather, it is to identify the options that can move the resources of the economy more in the direction how we can get more out of its resources within the current production possibility frontier and how the frontier can be expanded by technological and institutional development.

The Environmental (Physical) Damage Function

The environmental damage function describes the environmental damage due to environmental pressure from economic activities. The damage function underlying the marginal environmental burden to society shown in Figure 9.4 is a function including a term lifting the base variable by some exponent (e.g. Nordhaus and Boyer 2000) on climate change. Such a function possesses the mathematically convenient properties of being continuous, double differentiable and monotone. Scientists, however, have difficulties with recognising the working of nature in such damage functions. In nature you will typically expect this kind of simple relations only in rather narrow ranges delimited by discontinuities and qualitative shifts (e.g. Holling 1973, Muradian 2001, Walker and Meyers 2004). Ecosystems subject to environmental pressure can be

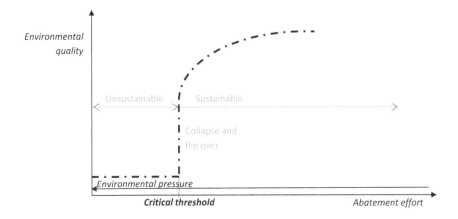

Figure 9.4   Environmental (physical) damage function

expected to shift between states, often with a considerable time delay ('hysteresis'), rather than responding with a small marginal change of environmental quality to small changes in environmental pressure irrespective of the level of environmental pressure. Often these shifts are irreversible.

Populations of fish, animals, etc. must exceed a minimum size to be able to reproduce. Global warming must be curbed to below 2°C to avoid runaway climate change. Contents of NO_X and SO_2 in precipitation must stay below the tolerance level of the vegetation to avoid loss of that particular form of life. Emissions of nutrients such as nitrogen and phosphorous beyond a threshold value leads to eutrophication and collapse of aquatic ecosystems.

It is difficult to boil down such complex functions to a two-dimensional framework, but the attempt is made in Figure 9.4 and Figure 9.5, inspired by earlier suggestions in the same direction (Perrings and Pearce 1994). They expose picture environmental quality as a function of environmental pressure, which again runs in the right-to-left direction.

The *marginal* damage function that follows from the relationship shown in Figure 9.4 is shown in Figure 9.5. The additional toll on environmental quality from one extra unit of environmental pressure increases as environmental pressure increases. At the threshold point, the marginal environmental damage caused by an additional unit of environmental pressure becomes very large as it becomes a matter of the *existence* of a set of ecological balances, species and ecosystems. Thus, the critical threshold marks the difference between 'sustainable' and 'unsustainable' if we by an 'environmentally sustainable economic activity' literally mean activity that can take place alongside the preservation or regeneration of the important environmental qualities.

Scientific knowledge is often not sufficient to identify exactly how much pressure the environment can take before collapsing. It is, however, often possible

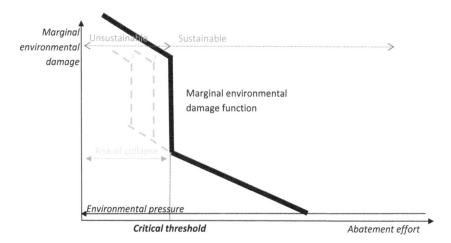

Figure 9.5   Marginal environmental (physical) damage function

to say with some degree of certainty at which point the risk of collapse appears. This point is essential for environmental regulation and planning.

On the safe side of the threshold, small changes in abatement will result in small changes in environmental quality whereas crossing the threshold triggers off a radical change.

The marginal damage in Figure 9.5 is measured in physical terms such as the deviation in fish population size from a natural, pristine or good ecological state or in some environmental quality index such as the deviation of global temperature from pre-industrial levels or the deviation of air quality from the air quality without serious impacts on human health or mortality. The efforts required to avoid it are, as shown in Figure 9.5, counted in euros or some other currency. To compare them it is necessary to compare the *value* of the damages with the value of the efforts required to avoid them. This step, the valuation of environmental damage, is undoubtedly the source of the biggest controversy in interdisciplinary analysis of environmentally sustainable development.

Environmental Valuation

The standard approach in environmental economics to environmental valuation is to apply the well-established framework of consumer theory. Consumer theory links the prices paid on the market to the utility of the consumers. The crucial assumptions are, first, that *utility is cardinal*, that is, can be measured on a continuous scale such that we can say that the utility of consuming a good is not only higher than the utility of consuming another good, but 2.5 times higher. In this way monetary values of the utilities from consuming various goods can be added for the individual consumer as well as across consumers. Second, all goods are *substitutable*, such that you can always replace a bundle of goods with an appropriate bundle of other goods and then get the same utility. These assumptions are necessary for the utility function to be *continuous* in the sense that a small change in the consumption of a good leads to a small change in utility. Third, since the transactions that take place on the market are voluntary, the market price represents the *willingness-to-pay (WTP)* of all the consumers for the marginal product. On this basis, the value of a change in consumption of a good for the consumers as a whole can be calculated. This value is usually understood as the value of the good to society or its *social value*. Moreover, it is directly comparable to the gross domestic product (GDP), the income or consumption opportunities of the economy as a whole.

One example of this line of reasoning is the (Nordhaus 1999) cost-benefit analyses of climate policy. Nordhaus accounted for all thinkable positive and negative effects including the negative effects on GDP of increased climate-related illnesses in Africa to GDP. It was calculated that 26 million years of life would be lost due to a 2.5°C global warming causing a loss of 4.6 per cent of GDP in Sub-saharan Africa. On the other hand, he also had found that outdoor summer leisure activities such as golfing was so valuable to Americans that a 2.5°C global warming would increase their utility corresponding to 0.3 per cent of GDP. The value of the extended summer season in the US amounts to $19 trillion whereas the value of the death and disease in Sub-saharan Africa amounted to $9 trillion.

There is intuitively something wrong with the assessment that the ability of Americans to extend the golf season with a few days should be more important to the global society than widespread disaster in Sub-saharan Africa. Even if it should be correct that the impact on GDP is as stated. Therefore, there is a long tradition in economics of pointing to such problems and devising alternatives to relying exclusively on consumer theory in environmental analysis.

To be fair, Nordhaus would probably agree that the international community should mobilise resources to combat malaria and other diseases in any case, but that there are cheaper ways to do it than stopping global warming and even that malaria should be eradicated altogether. But still, accounting of consumer values (measured in GDP changes) does not cover the social value of doing it.

Human beings have other value systems than those expressed by the consumer's WTP at the market place. They are expressed in ethical, moral and aesthetic norms and principles. Contrary to WTP, which is a purely individual concern, they are shared by groups' or cultures' people. They do play a role in the acts of consumers on the market place. They are often so limited that economic studies can ignore them without heavy loss of the applicability of the results.

In environmental policies or other policies, it is very different. Political priorities and their associated norms and principles shared by large groups or cultures are very important for the value of environmental damage to society.

There is a long tradition in economics to address this problem and devise alternatives. For instance (Sen 1977) criticised the omission of these other forms of rationality to result in a picture of man as 'rational fools' or 'social morons', also referred to as 'homo oeconomicus'. In philosophical contribution to the debate, (Sagoff 1998) suggests using different measures of our values when we act as citizens than when we act as consumers. Otherwise, we end up confusing the categories. In a recent report to an EU summit, Stiglitz, J.E., Sen, A. and Fitoussi, J-P. (2009) recommend that sustainable development should be described with a set of monetary *and* physical indicators. It is overly ambitious to attempt to convert all the dimensions of human well-being into a single metric. Not that there is anything wrong with the measure of GDP per se, but it cannot adequately measure all dimensions of social progress. In particular not the discontinuities and ethics involved in its environmental dimensions.

Consumer theory is perfectly well fitted to analyse the social value of what consumers carry home from the mall, but using it to quantify the social value of environmental goods is more problematic. The substitutability and continuity assumptions make sense in a mall where you can fill your trolley in thousands of different ways and still get the same utility out it. You can fill it in thousands of different ways that cost the same amount at the counter and you can find the consumption bundle that matches your preferences and your budget. In this world of myriad closely substitutable goods, it is always possible to express the value of one good in terms of others and, eventually, in a monetary value.

Some environmental qualities can also be put in the trolley in a special sense. Houses located next to the beach, the river or the forest are more expensive than otherwise comparable houses. This is because buying a property also buys privileged access to the environmental quality in question. Still, this is something different than the value to society of waters and forests that provide habitat to species and ecosystems and numerous other ecological functions.

Rational behaviour is based on a set of preferences or, more generally, intents, but they do not always adhere to the continuous utility function required by standard consumer theory. Some consumers would not substitute kosher or halal meat for pork at any price. There are also physical minimum levels of the amount of water, carbohydrates and protein we need to survive and no one can trade that away. None of us would consume human flesh, but on the other hand, if the alternative was to die, most of us probably would. This is, however, not a proof that 'everybody has his price', i.e., that everything is substitutable if the price is right. Rather it is a *dilemma*, which is something quite different from the normal economic choices to which economic theory applies.

When consumers prefer a given quantity of good or reject to consume any quantity of another good regardless of the prices involved, it is called *lexicographic preferences*. As noted by Malinvaud (1976) standard consumption theory is not compatible with such behaviour and they are usually excluded from economic analysis. There is no significant loss of generality from ignoring such behaviour, when analysing the market as a whole.

Recent studies have identified lexicographic preferences for environmental qualities along with the standard continuous-utility-preferences. Spash and Hanley (1995) and Spash (2000) operate with strong and weak modified lexicographic preferences. Scott (2002) looks for dominating preferences and Veisten, Navrud, and Valen (2006) found that 48 per cent had lexicographic preferences with respect to cutting of old forest, but only eight per cent to all the biodiversity-related questions in the survey. Such results point to the question of when environmental values should be addressed from the consumer perspective and when they should be addressed from the citizen perspective.

In public policy – not least environmental policy – strong principles that reject trade-offs are much more important than in consumer choices. Public policy choices are not only for the individual, but for all, including future generations. Thus they rely heavily on common norms that would appear in the analysis of consumers as lexicographic preferences. To the extent that lexicographic preferences really are citizen values, they may be better described as *political* or *societal priorities* and studied with methods appropriate for these kinds of values. Societal priorities are formed in the policy cycle and they are much more concerned with rights and minimum safe levels than willingness to pay for individual consumption. This includes the rights or entitlements of future generations to environmental qualities that the current generation has taken for granted, but nevertheless are destroying.

The principles of environmental policies are such as 'economic activity should not lead to further extinction of species'. 'People should not be exposed to life and health threatening chemicals or air quality in their daily life'. 'Runaway climate change should be prevented'. Cost is not an issue in these priorities. Even economic policies are loaded with principles such as: 'No new or higher taxes', 'we cannot afford the banks to go bankrupt', 'inflation rates will be kept below 2 per cent in Europe', etc. They all intend to sustain some set of balances or a system which is of a larger (but still unspecified) value than the immediate economic costs.

Environmental valuation studies usually make an account of consumers' willingness to pay for an environmental quality. It can be revealed from market behaviour or it can be stated in questionnaires. This approach seems adequate when studying the consumer value of a projected car make, but less so when studying the social value of an environmental programme. In the former case, the willingness to pay of the individual is of no concern to other consumers. The willingness to share environmental goods with future generations, however, does not make much sense unless a majority of citizens agree upon it and are willing to share the costs of doing it. The opinion of other citizens as well as the likelihood that all other citizens carry their fair share of the cost burden plays a crucial role in the formation of societal priorities. So do assessments of the likelihood of government actions to actually secure the environmental qualities for the future generations.

Thus, *group deliberation methods* are often better tools for stating political priorities than opinion polls or similar individual-based surveys. They are also used increasingly to inform political decisions. In Denmark, for instance, parliament established a Technological Council in 1995 which has used a range of group deliberation methods to analyse citizens' values on environmental and other issues. Political parties also make extensive use of focus groups.

In addition to the uncertainties of the physical damage function discussed above, monetary estimates of environmental values are surrounded by clods of uncertainty. Different methodologies that are equally scientifically recognised lead to different estimates. The same methodology used by different research teams can even lead to very different estimates. For instance, an important phenomenon observed in valuation studies is that the compensation required for accepting lower environmental quality ('willingness to accept') is much higher than willingness to pay (Knetsch and Sinden 1984, Knetsch 1989, Kahneman, D., Knetsch, J.L. and Thaler, R.H. 1991). This *endowment effect* probably has more to do with rights and entitlements than with preferences in the consumer universe.

In addition to this kind of methodological ambiguities and the statistical uncertainties of estimates, environmental valuation includes the fundamental uncertainties from the relation between environmental pressure and environmental damage. When environmental pressure exceeds the threshold beyond which the ecological balances may collapse as described above, we often face *scientific uncertainty* (or *ignorance*), which is different from risk (Knight 1921). Whereas risk or statistical uncertainty is based on experience where statistical patterns can be assumed to repeat themselves in the future, this knowledge is not there when we face scientific uncertainty. This is often the case for extreme environmental pressure and global warming as good example.

Figure 9.6 shows a cost function where the increasing marginal damage from no environmental pressure to the threshold at which important environmental qualities cannot be sustained. The uncertainty about the environmental pressure at which the irreversible losses of important environmental qualities will take place cannot be handled as if it was statistical uncertainty. If it was statistical uncertainty, the standard theory of expected utility would apply. Scientific uncertainty about irreversible losses of important qualities, however, leads to the precautionary principle, reflecting a prohibitively high environmental cost at a safe minimum standard in some safety distance from the point at which the loss can occur (Ciriacy-Wantrup 1952). Several routes in economic thinking have led to this result (Arrow and Hurwicz 1972, Woodward and Bishop 1997, Weitzman 2007).

Despite all these critical points the study of consumer values of environmental qualities is very important, but only for the study of consumer values. The study of citizen values requires methods designed for that purpose. Moreover, not all values can be reduced to one number. The alternative to this is multidimensional and interdisciplinary analysis of the problems and the solutions.

A Standard Model with Interdisciplinarily Consistency

Now, the revised components of the model of the environmental-economic balance can be put together again and this is done in Figure 9.7.

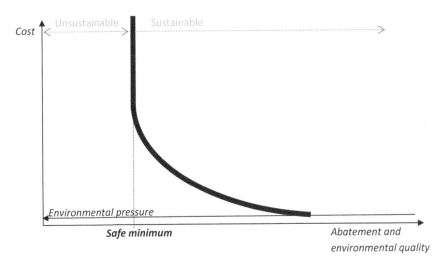

Figure 9.6   Sustainable and optimal environmental pressure

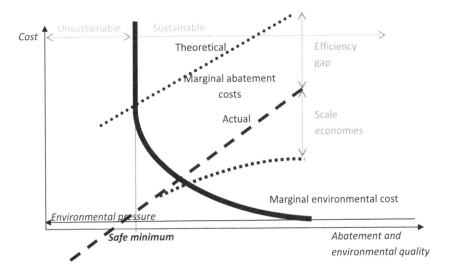

Figure 9.7   Marginal environmental cost

The theoretical as well as the actual marginal abatement cost curve can be situated in many ways. In Figure 9.7 the theoretical marginal abatement cost curve intersects the vertical segment of the marginal environmental cost curve, whereas the actual marginal abatement cost curve intersects at the declining segment. They could also cross at the vertical segment or both at the declining segment. Both will, however, intersect the marginal environmental cost curve at some point and this point will mark the optimal environmental pressure for the same reasons as in Figure 9.1. The fundamental condition for optimal environmental pressure, marginal abatement costs = marginal environmental costs, still holds. Only the curves are different.

If the intersection between the curves is at the vertical section of the marginal environmental cost curve, then it is very difficult if not impossible to know the economic value of environmental degradation. But it is also not necessary for establishing that the optimal environmental pressure is at the safe minimum. In mathematical terms, it is a corner solution. This means that the optimal environmental pressure in this case is the safe minimum, which is identifiable without any monetary estimate of the environmental value.

In that case the most important contribution from economics is about the transition: How fast to get there and by which technical and institutional changes. The time span of transition to environmentally sustainable economic activities is very important to the costs of transition. A fast transition must be expected to be more expensive than a longer transition due to premature scrapping of equipment and plants.

If the actual marginal abatement costs are low, that is, crossing at the declining section, an environmental pressure at its threshold and an environmental quality at its safe minimum standard are sub-optimal. The gains in well-being from a higher environmental standard may very well exceed the costs of reducing the environmental pressure. Reforestation close to urban areas for recreational reasons could be an example of such change. It is not – at least not initially – a matter of avoiding irreplaceable environmental loss, but just generally to improve our well-being.

Note, that in either case, there is no conflict between 'optimal' and 'sustainable' environmental pressure. An optimal environmental pressure is sustainable and that coincides with intuition: How could an unsustainable environmental pressure be optimal from the point of view of society?

Once the standard assumptions about the workings of nature and human values are the same in science, economics and philosophy, the sources of fundamental dispute disappears and interdisiplinary analysis can go from conflict to synergy. Thus, environmental-economic analyses of sustainable development that are multidimensional and interdisciplinary can be expected to get further than one-dimensional analyses with assumptions that are inconsistent with the standard assumptions in science, philosophy, psychology, etc.

The model can alternatively describe a model of a market for emission permits. In that case, the government would issue a number of emission permits corresponding to a reduced environmental pressure. The marginal abatement cost function will then be the demand function for polluters. They will abate environmental pressure (from the right) up to a point where it becomes cheaper to buy emission permits. If this price is lower than the social cost to be saved by further abatement, the government can reduce the number of permits and vice versa. Similar results can be obtained by taxing the environmental pressure.

Environmental change is usually long-term change. There are several examples of resolute efforts to abate water pollution in the 1970s that have led to a sustainable environmental pressure on the aquatic environment from urban waste water and lakes and a full recovery of lakes and streams. But it took about 30 years. Similarly, the transition to non-fossil and energy-efficient economic activities that will take place in the coming decade will pay off to the present generation as it saves expensive fossil fuels, but the pay-off in terms of avoiding runaway climate change, will only appear beyond the middle of this century.

The issue of discounting has been omitted from this chapter, to keep it short. It does, however, play a decisive role in the outcome of economic analyses of sustainability. We discount future economic values more, the further in the future they are. That is, 1,000 euros in 10 years are not as valuable to us in this year as 1,000 euros now and 1,000 euros in 20 years are even less valuable. The discount rate is a mathematical expression similar to the interest rate, to express how much we discount economic values in the future. The higher the rate of discount, the less future values are worth to us in the present.

Obviously, with such long-term planning horizons, where the expenses are now or in the near future and the benefits in the distant future, the result of any economic analysis is extremely sensitive to the discount rate assumed. On the other hand, the people living in the next century are not the same as the people living now. Nevertheless, the conditions they live under are determined by whether or not the people living today undertake the transitions to sustainable activities. Thus, there is inevitably an ethical aspect of such a long-term economic analysis. It would be unethical to let our decisions about transition guide by the discounting of future values, just because they belong to some other people.

Fortunately, recent advances in the economics of discounting show that the difference between maximising the present value of transition programmes and respecting fundamental principles in inter-generational ethics are not necessarily as large as previously assumed. Still, such long-term planning without inter-generational ethics would be incomplete and partial.

Three green waves in political opinion have rolled over Europe (and most of the World) in the early 1970s, in the mid-1980s and in the late 2000s. Obviously, if the public opinion matters to the value of something to the society, the value of the natural environment has become larger to society after each wave than it was before. There is little reason to believe that environmental qualities will be valued the same way in, say, 40 years' time as they are now. The impacts on environmental quality of the environmental decisions taken today often have that kind of lead time.

There are several good attempts in climate change economics such as the IPCC reports and the Stern Review (Stern 2007) to achieve interdisciplinary consistency. It is definitely not the only way to do it, but it clearly demonstrates the importance of an interdisciplinary and multidimensional approach to such analyses.

As it appears from the adapted environmental-economic model, the concepts of 'optimal environmental pressure' and 'sustainable development' are not necessarily different after all. It is all in the assumptions.

References

Arrow, K.J. and Hurwicz, L. 1972. An Optimality Criterion for Decision-making Under Ignorance, in *Uncertainty and Expectations in Economics: Essays in Honour of G.L.S. Shackle*, edited by C. Carter and J.L. Ford. Oxford: Basil Blackwell.

Atkinson, G., Dietz, S. and Neumayer, E. 2007. *Handbook of Sustainable Development*. Cheltenham: Edward Elgar Publishing.

Ciriacy-Wantrup, S.V. 1952. *Resource Conservation: Economics and Policy.* Berkeley, CA: University of California Press.

Council of the European Union. 2006. *Renewed EU Sustainable Development Strategy*. 10917/06.

Daly, H.E. 1992. Is the Entropy Law Relevant to the Economics of Natural Resource Scarcity? Yes, of Course it Is! *Journal of Environmental Economics and Management*, 23, 91–5.

Daly, H.E. 1997. Georgescu-Roegen Versus Solow/Stiglitz. *Ecological Economics*, 22, 261–73.

Dasgupta, P. and Heal, G. 1974. The Optimal Depletion of Exhaustible Resources. *The Review of Economic Studies*, 41, 3–28.

De Groot, R., Van der Perk, J., Chiesura, A. and van Vliet, A. 2003. Importance and Threat as Determining Factors for Criticality of Natural Capital. *Ecological Economics*, 44, 187–204.

Ekins, P. 2003. Identifying Critical Natural Capital: Conclusions About Critical Natural Capital. *Ecological Economics*, 44, 277–92.

Ekins, P., Folke, C. and De Groot, R. 2003. Identifying Critical Natural Capital. *Ecological Economics*, 44, 159–63.

Ekins, P. and Simon, S. 2003. An Illustrative Application of the Critic Framework to the UK. *Ecological Economics*, 44, 255–75.

Ekins, P., Simon, S., Deutsch, L., Folke, C. and De Groot, R. 2003. A Framework for the Practical Application of the Concepts of Critical Natural Capital and Strong Sustainability. *Ecological Economics*, 44, 165–85.

Finansdepartementet. 2001. *Retningslinjer for Den Økonomiske Politikken*, St.meld. nr. 29 (2000–2001), [Online]. Available at: http://www.regjeringen. no/nb/dep/fin/dok/regpubl/stmeld/20002001/stmeld-nr-29-2000-2001- .html?id=194346 [accessed: 22 February 2010].

Georgescu-Roegen, N. 1971. *The Entropy Law and the Economic Process*. Cambridge, MA: Harvard University Press.

Goeller, H.E. and Weinberg, A.M. 1976. The Age of Substitutability. *Science*, 191, 683–9.

Hansen, A.C. 2001. Estimating Non-renewable Resource Capital Consumption, in *The Sustainability of Long-Term Growth*, edited by M. Munashinghe, O. Sunkel and C.D. Miguel. Cheltenham: Edward Elgar Publishing, 397–421.

Hansen, A.C. 2002a. Bæredygtig Økonomi – 10 År Efter Rio. *Miljøforskning*, 16–21.

Hansen, A.C. 2002b. *What Happened to Sustainable Savings?* [Online]. Research Paper Series 2002/16. Department of Social Science at Roskilde University. Available at: http://www.ssc.ruc.dk/workingpapers/2002/16-02.pdf [accessed: 9 November 2009].

Hartwick, J.M. 1977. Intergenerational Equity and the Investing of Rents from Exhaustible Resources. *The American Economic Review*, 67, 972–4.

Hartwick, J.M. 1994. National Wealth and Net National Product. *The Scandinavian Journal of Economics*, 96, 253–6.

Holling, C.S. 1973. Resilience and Stability of Ecological Systems. *Annual Review of Ecology and Systematics*, 4, 1–23.

Hourcade, J.-C. and Robinson, J. 1996. Mitigating Factors: Assessing the Costs of Reducing Ghg Emissions. *Energy Policy*, 24, 863–73.

Intergovernmental Panel on Climate Change, IPCC. 1996. *Climate Change 1995: Economic and Social Dimensions of Climate Change*. New York: Cambridge University Press.

Intergovernmental Panel on Climate Change, IPCC. 2001. *Climate Change 2001: Mitigation*. New York: Cambridge University Press.

International Energy Agency, IEA. 2008. *World Energy Outlook 2008*. Paris: IEA.

Jaffe, A.B. and Stavins, R.N. 1994a. The Energy Paradox and the Diffusion of Conservation Technology. *Resource and Energy Economics*, 16, 91–122.

Jaffe, A.B. and Stavins, R.N. 1994b. The Energy-Efficiency Gap What Does it Mean? *Energy Policy*, 22, 804–10.

Kahneman, D., Knetsch, J.L. and Thaler, R.H. 1991. Anomalies: The Endowment Effect, Loss Aversion, and Status Quo Bias. *The Journal of Economic Perspectives*, 5, 193–206.

Knetsch, J.L. 1989. The Endowment Effect and Evidence of Nonreversible Indifference Curves. *The American Economic Review*, 79, 1277–84.

Knetsch, J.L. and Sinden, J.A. 1984. Willingness to Pay and Compensation Demanded: Experimental Evidence of an Unexpected Disparity in Measures of Value. *The Quarterly Journal of Economics*, 99, 507–21.

Knight, F.H. 1921. *Risk, Uncertainty, and Profit*. Hart, Schaffner & Marx Prize Essays, No. 31 Boston and New York: Houghton Mifflin.

Malinvaud, E. 1976. Lectures on Microeconomic Theory. Amsterdam: North-Holland.

Meadows, D.H., Meadows, D.L. and Randers, J.B.I. 1972. *The Limits to Growth: A Report for the Club of Rome's Project on the Predicament of Mankind*. London: Earth Island.

Medvedev, D. 2009. Go Russia!, in *President of Russia Website*. [Online]. Available at: http://eng.kremlin.ru/speeches/2009/09/10/1534_type104017_221527.shtml [accessed: 22 February 2010].

Muradian, R. 2001. Ecological Thresholds: A Survey. *Ecological Economics*, 38, 7–24.

Neumayer, E. 2003. *Weak Versus Strong Sustainability: Exploring the Limits of Two Opposing Paradigms*. Cheltenham: Edward Elgar.

Nordhaus, W.D. and Boyer, J. 2000. *Warming the World: Economic Models of Global Warming*. Cambridge, MA: The MIT Press.

Nordhaus, W.D. 1999. *Roll the Dice Again: Economic Models of Global Warming*. [Online]. Available at: http://www.econ.yale.edu/~nordhaus/homepage/dicemodels.htm [accessed: 25 October 2009].

Organisation for Economic Co-operation and Development, OECD, and International Energy Agency, IEA. 2000. *Experience Curves for Energy Technology Policy*. Paris: OECD/IEA.

Pearce, D.W. and Turner, R.K. 1991. *Economics of Natural Resources and the Environment*. Baltimore: Johns Hopkins University Press.

Perrings, C. and Pearce, D. 1994. Threshold Effects and Incentives for the Conservation of Biodiversity. *Environmental & Resource Economics*, 4, 13–28.

Porter, M.E. and Linde, C.v.d. 1995. Toward a New Conception of the Environment-Competitiveness Relationship. *The Journal of Economic Perspectives*, 9, 97–118.

Sagoff, M. 1998. Aggregation and Deliberation in Valuing Environmental Public Goods: A Look Beyond Contingent Pricing. *Ecological Economics*, 24, 213–30.

Scott, A. 2002. Identifying and Analysing Dominant Preferences in Discrete Choice Experiments: An Application in Health Care. *Journal of Economic Psychology*, 23, 383–98.

Sen, A.K. 1977. Rational Fools – Critique of Behavioural Foundations of Economic-Theory. *Philosophy & Public Affairs*, 6, 317–44.

Solow, R. 1974. Intergenerational Equity and Exhaustible Resources. *Review of Economic Studies*, 41, 29–45.

Spash, C.L. 2000. Multiple Value Expression in Contingent Valuation: Economics and Ethics. *Environmental Science & Technology*, 34, 1433–38.

Spash, C.L. and Hanley, N. 1995. Preferences, Information and Biodiversity Preservation. *Ecological Economics*, 12, 191–208.

Stern, N. 2007. *The Economics of Climate Change: The Stern Review*. New York: Cambridge University Press.

Stiglitz, J. 1974. Growth with Exhaustible Natural Resources: Efficient and Optimal Growth Paths. *The Review of Economic Studies*, 41, 123–37.

Stiglitz, J.E., Sen, A. and Fitoussi, J.-P. 2009. *Report by the Commission on the Measurement of Economic Performance and Social Progress*. [Online]. Available at: www.stiglitz-sen-fitoussi.fr. [accessed: 9 November 2009].

United Nations Conference on Environment and Development, UNCED. 1992. *The Rio Declaration on Environment and Development*. United Nations Conference on Environment and Development.

Veisten, K., Navrud, S. and Valen, J.S.Y. 2006. Lexicographic Preference in Biodiversity Valuation: Tests of Inconsistencies and Willingness to Pay. *Journal of Environmental Planning and Management*, 49, 167–80.

Walker, B. and Meyers, J.A. 2004. Thresholds in Ecological and Social–Ecological Systems: A Developing Database. *Ecology and Society*, 9(2): 3. [Online]. Available at: http://ww.ecologyandsociety.org/vol9/iss2/art3/ [accessed: 22 February 2010].

Weitzman, M.L. 2007. A Review of the Stern Review on the Economics of Climate Change. *Journal of Economic Literature*, 45, 703–24.

Woodward, R.T. and Bishop, R.C. 1997. How to Decide When Experts Disagree: Uncertainty-Based Choice Rules in Environmental Policy. *Land Economics*, 73, 492–507.

World Commission on Environment and Development, WCED. 1987. *Our Common Future*. Oxford: Oxford University Press.

PART III

Sustainability in an Everyday Life Perspective

Local Experimentation and Deliberation for Sustainable Development: Local Agenda 21 Governance

Jesper Holm

The 1992 Rio process focused on enabling a sustainable agenda for the twenty-first century under the auspices of the United Nations. It encountered a new multilayered and governance-based regulatory regime, a regime that Martin Jänicke has labelled the Rio model of environmental governance (Jänicke 2006). This regime of voluntary approaches to environmental policy innovations, lesson-drawing and policy diffusion is very much a knowledge-driven effort to stimulate commitment to sustainable development. The regime is characterised by long-term goals, sector integration, stakeholder participation and activated self-regulation. Our main interest in this chapter is the challenges to this type of governance for *local* transition processes of our consumption patterns, production and mobility towards sustainability.

One well-known part of the Rio process developed along the line of *ecological modernisation* in the late 1980s and the 1990s. Environmental policy related to industrial manufacturing and products came to adopt this new development path and became 'eco-modernistic', interactive and de-formalised (Jänicke 2000, Holm 1999, Remmen 2006). However, in many European countries, as in Scandinavia, the local authorities have to a varying degree failed in the firm implementation and enforcement of this new regulatory paradigm, as it does not fit the hitherto legal control policy style in local environmental policy (Mortensen 2001, Holm, Hansen and Søndergaard 2003, Jørgensen and Lauridsen 2007).

Another window of opportunity for the local authorities to find a political role in the Rio model was to engage in Local Agenda 21 (LA21). Local Agenda 21 was addressed in Chapter 28 of the AGENDA 21 document (United Nation's Programme of Action for Sustainable Development) to bridge civic servants, citizens and NGOs in consensus-oriented deliberation for environmental initiatives. It appealed to all governments and local authorities of the member states of the United Nation to draw up local sustainable development action plans for the twenty-first Century. The document claimed a call for greater public participation at the local level: 'As the level of governance closest to the people, they play a vital role in educating, mobilising and responding to the public to promote sustainable development' (United Nations 1993).

In the document, attention was devoted to the serious involvement of local authorities, corporations and citizens in the planning of the common good (United Nations, 1994). A spirit of deliberative commitment dominated where all actors are supposed to have a common interest in co-operating for sustainable development. A dialogue process with citizens, local organisations and private enterprises forms the basis for an LA21 as a shared community image that takes place by '... consultation and consensus-building ... where local authorities learn and acquire the information needed to form the best strategies' (Ibid.). To foster and encourage a more sustainable development depends thus upon something more than the hitherto normal environmental policy or ecological modernisation. The political deliberation was and is inspired by aspiration for substantial changes:

> (...) sustainable development will require thorough changes of society – from basic infrastructure to households (...) But the aim is to go beyond simple hearings... it is also an aim to involve citizens in order to elevate the role of the municipalities from being authorities to becoming supervisors and partners. (Miljøministeriet 1995 [Danish Ministry of Environment])

This pre-assumption in the Rio model about a deliberative commitment was also behind the LA21 implementation style in the Scandinavian countries, which had and have a relatively firm institutional capacity on a local or regional level when it comes to environmental infrastructure, licensing, enforcing national standards, etc. (Lafferty 1999). There have been quite impressive and intensive campaigns, strategic programmes and sector initiatives under the LA21 umbrella for local action on sustainable development in Scandinavia since 1994 (for an overview, see Lafferty 2001, Norland, Bjørnæs and Coenen 2003, Bjørnæs et al. 2004). The Nordic countries are, according to some studies and surveys, thought to be stronger on LA21 politics than the rest of Europe. Thus, the endeavours in these countries may serve as an extreme case to test and discuss LA21 in the Western part of the world. To be sure, based on national surveys and case studies we have also maintained a general critique of the lack of continuity in the LA21 process. There has been far too much window dressing; too low ambitions in community transition; no change of basic environmental, business and welfare policy; and very few attempts at fundamental participatory approaches except from Denmark (Bjørnæs et al. 2004, Holm 2004). Even though the critics of blurred environmental impacts and window dressing among the manifold efforts in Scandinavia are important, they will not be repeated here (see, e.g. Lafferty 2001, Norland, Bjørnæs and Coenen 2003).

In this chapter we elaborate upon the interesting fact that inherent in the manifold LA21 policy documents and initiatives lies a new understanding of the necessity to break with the barriers and path dependencies that occur in normal environmental policies, to perform rule-altering sub politics (Beck 1999), to construct niche initiatives beside markets and search for value-based politics. Thus, we here want to frame LA21 as societal experimentation processes of participative deliberation, network governance and niche experiments with the aim of developing *more* than ecological modernisation: what others call structural change (Jänicke 2000) or environmental transition (Kemp and Rotmans 2001). This chapter offers an initial focus on this from a theoretical position and partly brings in what kind of interesting paradigmatic openings we are to find in Scandinavia and more specifically Denmark, one of the pioneering European countries when it comes to LA21.

We start the search by establishing a theoretical framework for evaluating the deliberative and experimental policy style of LA21. We briefly discuss environmental transition and the role of politics to identify governance options for a comprehensive and radical environmental transition in local manufacturing, housing, consumption and mobility – in short, sustaining socio-technical systems. The current challenges for a *local* transition towards sustainability are called upon under these theoretically labelled development challenges of governance, by focusing on niche development, experimentation and transition management. Secondly, a short presentation of reflexive and political modernisation (as developed by Beck (1994) and Hajer (1996)) is introduced in order to elaborate on the forming of LA21 political agencies fostering the announced deliberation and transition. The presentation serves to frame the mode of operation, function and agency of new local policy styles fostered within the predominant LA21 regime in Scandinavia. With a background knowledge in national and local studies (surveys, in-depth cases and policy analysis) of LA21 policies and activities in Denmark, the chapter ends by discussing some of the outcomes of LA21 policy measures in order to discuss societal learning and options for enhancing sustainable development paths through the new types of political modernisation and transition governance.

Environmental Transition

Let us look at the local community targets of LA21, such as the initial Danish LA21 policy principles of cross-sectoral and holistic efforts, active public participation, a cradle-to-grave perspective, global concern, and a long time span perspective. This has to do with sustainable ways of producing, eating, housing, getting around and living. Of course this implies profound changes in our present practices – changes which cannot be restricted to environmental improvement or the optimisation of present systems of production and consumption, but have to include radical system innovations. Thus, we must pay attention to how we may initiate and implement the development of more sustainable consumption patterns, socio-technical systems and community coherence (Weaver et al. 2000, Kemp 2003). Again, taking for granted that hitherto regulation of pollutants as thresholds standards, environmental standards, green management schemes, etc. only serve to clean up and *not* perform sustainable development paths.

A way of comprehending such major developments inspired by economics relates to transition theory, as an interventionist approach looking at how we may exercise the governance of systemic sustainable change processes (Kemp and Loorbach

2003). From this point of view, technological, structural and cultural changes are highly related, forming a seamless web. For example, any change towards more sustainable mobility patterns will include new technologies interacting with social and cultural changes. The position taken is that any transition towards sustainability will imply a high level of social-cultural change coupled to a similarly high level of technological change. This is also called system innovation:

> System innovations are defined as major changes in the way societal functions such as transportation, communication, energy supply; feeding, housing and water management are fulfilled. Such major changes typically involve a co-evolution of a number of related elements, including technology, infrastructures, symbolic meanings, regulation, scientific knowledge, industry structures, etc. (Kemp, Weber and Schot 2001)

The aims of LA21 could well be described as a transition to a more sustainable local society and imply, according to this school of thought, a structural change involving the adoption and diffusion of new technologies and consumption patterns embedded in new economic, social, institutional and cultural relations. Such changes take place at the level of systems of production and services, distribution and consumption and are thus the result of system innovations. System innovations involve a wide range of actors, including firms, consumers, citizens, NGOs, knowledge producers and governments. They are not caused by a change in a single factor but are the result of the interplay of many factors and actors that influence each other. They imply change at various levels: at the micro-level of individual actions, at the meso-level of structuring paradigms and rules and at the macro-level comprising a deep structural level of trends such as individualisation and globalisation.

We may identify an array of transition approaches which are oriented towards system innovations with long-term visions which call for radical shifts, but envision the process as an evolutionary journey consisting of small incremental steps as suggested by Vellinga (2000) and Kemp and Loorbach (2003). They share a common understanding of the process as concurrent changes in environmental, economic, technological, social and cultural dimensions – a process which is modulated in governance among networks, of co-development among producers, customers, citizens and officials, of incentives structures and of 'institutions' capable of generating and stabilising visionary perspective of the process.

A dominant approach in modelling and conceptualising transitions processes has been the 'three-layer model', with socio-technological regimes operating within relatively stable landscapes (national and global conditions) and in a dynamic interplay with niches of alternative technologies and practices, where new openings and possibilities are created (Kemp, Schot and Hoggma 1998, Geels 2002, Kemp and Loorbach 2003, Smith 2003).

The concept of socio-technological regimes derives from the understanding of technology (manufacturing, mobility, infrastructure, consumption products, etc.) as being embedded in socio-technical combinations of elements such as technology/artefacts, knowledge, production structures, infrastructure, regulation, policies, practice, symbolic meaning, linked and aligned (Geels 2002, Rip and Kemp 1998). The elements and the linkages are the result of activities of social groups which (re)produce them (Geels 2002). Such socio-technical configurations also imply the sedimentation of specific technological trajectories, that is, they can be seen as technological regimes having stability and guiding innovative activities. Rip and Kemp (1998) state that:

> A technological regime is the rule-set of grammar embedded in a complex of engineering practices, production process technologies, product characteristics, skills and procedures, ways of handling relevant artefacts and persons, ways of defining problems; all of them embedded in institutions and infrastructures.

By this extension of the understanding of regimes – paths of technology development are shaped in an complex interplay of different actors; not only communities of engineers but also producers, users, social groups/stakeholders, policy makers/public authorities, financial networks, research networks, etc.

Socio-technical regimes work and develop within socio-technical landscapes, an external structure or context for the interaction of actors. The term landscape connotes a higher degree of stability, which may be founded in such different elements as price structures, political coalitions, dominating norms and values, environmental problems, and we may add local community structures, etc. (Geels 2002) – all elements that may be considered as external to the processes of the regime. Changes in the socio-technical landscape can stress regimes, for example the pressure for radical changes put on the energy regime by the global climate problem.

Regimes frame incremental development along their technological trajectories, whereas radical innovations are generated in niches (Geels 2002). Such niches may evolve on the fringe of or independently of the dominating regimes (Smith, Berghout and Stirling 2003), be the object of strategic management (Kemp, Schot and Hoggma 1998) and be supported and developed, e.g. by social experimentation (Brown et al. 2003) or the creation of technological niches (Truffer, Metzner and Hoogma 2002). Within this conceptualisation 'niches', understood as protected fields, are assigned a pivotal role in transition processes. Again, the fate of such niches for socio-technological systems is determined in interplay with the dynamic development within the regimes and at the level of socio-technical landscapes.

In an evolutionary perspective, niches create variation. The problem, however, is how transition processes may be supported given the selection exerted by dominant regimes.

The Governance of Environmental Transition Processes

As we have learned from the cultural politics tradition, the challenge concerns whether or not we may exercise the deliberate and purposeful governance of transition processes on top of social and economic acceptance? Part of this problem is at which level (households, projects, alternative technologies, institutions, industries, transition fields, technology systems, communities...) we may install such processes, and the problem of agency. Local Agenda 21 documents and strategies in Scandinavia address arenas such as public purchasing, construction, greening waste handling, public transport, school schemes and labels like the blue flag for tourism, etc., whereas the agencies addressed are typically entrepreneurial *individuals*. Studies of transition processes (toward more sustainable practices), however, have to an increasing extent taken systemic approaches, making the question of agency a question of constituency processes of *networks and coalitions of actors*. New policy schemes of transition (transition management, industrial transformation, (...) address the systemic level more directly – system innovation and transition, changes of socio-technological regimes are the direct perspective (Kemp and Rothmans 2001, Kemp and Loorbach 2003, Rotmans 2003). This involves policy means that can support the interplay of systemic changes and projects/technologies/niche markets (Kemp, Schot and Hoggma 1998, Smith 2003) in an orchestrated way and as part of this are capable of developing collective visions and goals (Weaver et al. 2000, Quist and Vergragt 2003).

These and parallel studies of more radical innovations make the configuration of specific, motivated actors a crucial step in the process of establishing new paths within a regime. The question of agency concerns, however, who can, and by which means, shape and constitute such networks, and the question of what we know about how such networks come about in processes of more radical environmental innovations.

The interplay of viewing and doing has been central in so-called transition management. Transition management, mainly developed in relation to Dutch programmes of sustainability, has addressed the problem of installing and managing transition dynamics within regimes, not only to optimise existing systems but to transform them into more sustainable patterns of production and consumption (Weaver et al. 2000, Kemp and Rotmans 2001, Grin, Rotmans and Schot 2009). They are strategic – as they have a reference to a perspective of sustainability. Transition management intends to combine an effort of short-term incremental improvements, using ongoing dynamics in the regimes, with a long-term horizon where visions and targets guide the overall transition process. The means are described as e.g. future scenario tools, technology portfolio-management and process management, but there must be many other ways. The policy style is one of deliberative efforts to ensure variety in technology options, learning and adaptability in both policies and short-term projects and, in this way, also avoid a lock-in to sub-optimal systems in the process (Rennings et al. 2003). This includes the organising of a multi-actor network which can elaborate and negotiate goals and visions. Furthermore, transition management requests/demands the mobilisation of a wider group of actors capable of carrying through experiments and programmes of innovations. A crucial activity in this way is the creation of what has been labelled 'transition arenas':

In it first phase, the transition arena is a relatively small network of innovators and strategic thinkers from different backgrounds that discusses the transition problem integrally and outlines the transition goals. In this phase, it is important to come up with creative, inspiring and integrating goals. (Kemp and Loorbach 2003)

When we deal with LA21, the efforts to govern current socio-technical systems for sustainable transition imply the problems of scale; as to what degree local agencies and networks have any possibility to alter existing technology systems as they are normally embedded in regional, national or global chains, and clusters, etc. One option is to narrow the transition policies to experiments, and niche management which, as we shall see, is also the case of the majority of LA21 efforts. Three major problems for LA21 niche policies are then:

- Which kind of arenas should be addressed attention and initiatives at a local level?
- Which kind of actor network can be configured?
- How do we ensure that successful niche experiments diffuse and seed transformations in technological regimes when LA21 efforts by definition are local?

In LA21 policies we are dealing with, first of all, local environmental authorities in deliberative efforts concerning changing patterns of consumption, mobility, housing, services, heating, energy use, etc.:

The lower authorities, such as provincial and municipal authorities, also have a role to play in transition management. They stand closer to citizens, than the national government, the local situation can permit radical experiments (such as car-free town centre or city heating). Furthermore, they have been assigned their own tasks in areas that are often relevant with regard to societal transformations, such as environmental planning, house construction, the environment and waste. (Rotmans 2001)

Rotmans (2001) points to experimentation on a small scale as optional for local, environmental transition politics, where local authorities may use their regulatory jurisdiction to frame, coach and govern transition management.

LA21, Niches and Transition

Within the niche management perspective, special attention is paid to the very early stages of technology development – in particular in relation to more radical innovations. Creating technological niches is considered as an important instrument. Such 'technological niches' are:

(...) protected spaces created by specific actors – by the industry managers, policy makers or citizen groups – with the strategic aim to test and develop a technology and to prepare it for further diffusion. (Truffer, Metzner and Hoogma 2002: 113)

Technological niches should be differentiated from market niches. In the case of market niches, a certain number of users recognise the advantages of the product or technology or the related image without difficulty, and that is why companies start producing it. In the case of technological niches, the user-specific advantages are less clear, so that constituents who are motivated by expected advantages apply temporary protection measures to interest the potential users in trying out the product (Truffer, Metzner and Hoogma 2002: 113). The means can be real-world experiments of forming new infrastructure, food and consumption habits, etc. and the corresponding use of new technological devices. The main objectives are: (1) to learn about the technology and factors of importance in real-life environments (competence building); and (2) to facilitate a societal embedding. Truffer, Metzner and Hoogma (2002: 113) point to three main processes of such a social embedding: (1) network management: the creation of a constituency for the technology or product (potential producers, users and/or regulators of the product who are willing to support the development of the product); (2) infrastructure matching: the fit of the new technology into an existing socio-technical environment (defined by physical infrastructure, legal settings, product standards and so on); (3) expectation building: the creation of positive expectations about the prospects and benefits of the new technology-service. In relation to alternative (maybe radically new) technologies, niches are the locus of learning and development processes which in turn may give impetus to transforming a socio-technical system:

Alternative technologies are developed at the micro level, particularly in these niches, where there are learning processes with regard to innovations, new practices or behaviour. As a result of these niches, options can be developed from ideas to alternatives. (...) learning process concerning alternative options and the forming of new actor networks in niches can produce bottom-up changes to regimes. (Rotmans 2001)

Jacobsen and Bergek (2003) point to the building of social networks and development of capacities in terms of knowledge, the linking of resources and formation of markets as the vital phases, and a deliberate building of coalitions of actors is identified as a critical condition.

What actually may enhance local communities to perform niche experiments and the forming of alternative technologies and consumption patterns is somewhat unclear – catalysts, learning processes (Rotmanns 2001), tensions in existing regimes as a result of changes in the landscape are mentioned (Smith 2003). But one of the biggest problems for transitions is the long time period over which these take place. Under normal circumstances, transitions require two to three decades.

If the preparation phase of experimentation and development of new technology is included in this, the transition period may take 40 to 50 years!

Leaving the issue of the generic and systemic transition needs, we will discuss how agencies and cognitive frames may be raised to subscribe for transition arenas. We do this by a reflexive institutional and cultural politics approach to the current ecological modernisation development, as it may enhance development dynamics and identities, which governed under the right conditions would foster changed institutions and ecological deliberation.

Reflexivity and LA21

Many efforts at comprehensive change on a community scale emerge under the conditions of reflexive, political modernisation among governed policy networks, as suggested by Beck (1994) and by Held (1995). The politically modernised state is, according to this approach, best equipped to act as a responsive or inter-active state and plays a very indirect or intermediary role in ecological transition. Such a state should orient itself to merely formulating general environmental standards and scenarios to spur self-regulation and deliberation, thereby leaving it to business organisations, NGOs and citizens' groups to make environmental initiatives. In other words, LA21 politics! The policy-style focus is upon voluntary and communicative measures, stressing the need to incorporate stakeholders and target groups for consultation, co-production of collective images and negotiating agreements on performance and measures. In Beck's terms: weakening rule-directed and strengthening rule-altering arrangements (Beck 1994: 40–2). Geus (1996) finds that environmental problems are to be approached by the state via the creation of situations and conditions that will make it attractive to make environmentally positive choices. It is regulation through a contextual or procedural shift, rather than the more traditional coercive measures. These perspectives on regulation build upon the theorem of institutional reflexivity related to the environmental policy area (Murphy 2000). This is described as a process in which modern society and its institutions attach specific meaning to the environmental problems, and build up capability to reflect on the social causalities to environmental problems and to perform adequate response (Beck 1994). It is a central part of this understanding that in this reflexive process, both the institutions and ways of reflection are changed. Changes occur in the roles and relationships of industries, government, citizens and agents (such as NGOs). This places the focus on how the environment is perceived and how it is communicated, on which agents are involved and how different agents reflect, act and communicate environmentally. Taking a departure in institutional reflexivity, environmental transition does not come about merely as a result of the anticipation of economic profit or obeying rules. But what then makes the reflexive process develop? What then become meaningful triggers among the interacting/communicating parties?

Cultural Politics and Identities for Environmental Changes

In Marteen Hajer's (1996, 2001) way of describing ongoing interpretations of the dynamics behind environmental changes, we find a reflective-constructivist approach that provides some interesting ideas concerning these kinds of questions when it comes to deliberation in LA21. Hajer's perspective of cultural politics pays attention to social narratives, storylines as compelling social constructions of the environment, nature and society that form institutions and policies, providing preferences and privileges for some coalitions, while suppressing others. In this 'cultural politics' approach, the focus is on the social order and the social impact of constructing technical, cultural or scientific definitions of environmental problems. From this point of view, current environmental policy does not move towards a transition as the storylines often maintain an expert-laity dichotomy, the predominant consumption-production pattern and the alienation of the lifeworld experiences of social need and struggle in the handling of the environment as an externality. The cultural politics approach also has a normative standpoint that wants to leave appropriate responses to particular 'environmental problems' open for local solutions, in a bottom-up process of articulating cultural values and social needs, whereby environmental issues from a thousand lifeworlds may be re-articulated.

The benefit of the focus on cultural politics in this context is that it fits so well with the LA21 policy style, as it basically relates to an embedding of normative claims and rationalities within a civic society, where everyday life experiences and values are coherently present in preferences and practice (see, e.g. Lash, Giddens and Wynne 1996). The focus is on diversity, plurality, temporality and situations that may produce a number of responses embedding environmental issues into local cultural politics of the good life, but also to (a hope for) a critique of growth, consumerism, etc. (Hajer and Fischer 2000). Thus, cultural politics focuses on how individuals, groups and discourse coalitions form interpretations of situations, conflicts and norms by the active shaping of narrative meaning – storylines. This places a focus on how the environment is perceived and how it is communicated, the agents that are involved, and how those agents reflect, act and communicate environmentally. In Hajer's (1996) terms, the cultural politics of, for example, LA21 could positively form discourse coalitions and storylines that enter the institutional level of sectors and gradually become a new institutional field. But these politics will have to rely on culturally, everyday life-based interpretations of environmental issues in their socio-economic and cultural meaning, if they are to achieve social mobilisation and political legitimacy. These institutions both deliberately and accidently co-shape the identity of community life in social and health institutions, the local trade and the public sphere and provide a spur to reflexive reactions among business and institutions, e.g. incorporating environmental profiles as the new normality. The institutionalisation of cultural politics is subsequently being diffused (in social fora, markets and commodities). Hajer would surely maintain that in order to be democratic the institutionalisation would have to be open for local social productive interpretations of environmental issues.

At stake are bottom-up processes of articulating cultural values and social needs, whereby environmental issues from a thousand lifeworlds may be re-articulated. The local community may be a promising starting point for this process of including phenomena of importance to everyday life, and also for making a link to technological systems for experimentation. Thus, ecological transition cannot be based on government intervention or corporate greening alone, but involves social entrepreneurs in a process of projects, future scenario efforts, etc. that may allow social change to take place democratically and based partially on a shared vision of the future. But the effort to perform a pragmatic yet visionary imagination of a wanted future must be followed by an effort to schedule the time-based step-by-step governance of a sector, a technological system or an area. Thus, deliberation and experimentation must include strategic niche management to perform transition.

Scandinavia, Denmark and LA21: Experimentation, Deliberation and Transition Options

The presumption about a deliberative commitment also seems to lie behind the implementation of LA21 in Scandinavian countries since 1995. All Scandinavian countries had and have a relatively firm institutional capacity on local or regional levels when it comes to environmental infrastructure, licensing, enforcing national standards, etc. In other countries (including Norway in part), the LA21 process has joined-in general initiatives for breaking with central politico-administrative domination, so as to build the local environmental administration and governments from the ground upwards (Lafferty 1999). The political deliberation was and is inspired by the aspiration for big changes. From national surveys and case studies I have been impressed by the level of activities being raised among municipalities, but also maintained a general critique of the lack of continuity in the LA21 process: far too much window dressing; too unambitious in terms of community transition; no change of basic environmental, business and welfare policy; and very few attempts to introduce fundamental participatory approaches except from Denmark (Bjørnæs et al. 2004). It is no surprise that the dominant impression is that the transformative impacts on core LA21 issues such as environmental sector coordination, transforming consumerism, the scale of societal metabolism with nature, democratic participation, cleaner technology systems and services are modest (Bjørnæs et al. 2004).

The LA21 process and initiatives have turned out differently among the Scandinavian countries due to different policy styles, politico-administrative structures and local cultures. Sweden has generated state-driven large investment programmes (LIP) for local development in energy-saving measures and environmentally benign infrastructure, whereas new 'practical activism' among NGOs and green entrepreneurs in many residential areas was initiated more strongly in Denmark and Finland. In Norway LA21 contributed to a period of the decentralisation of hitherto national environmental policy initiatives and administration.

The Historical Development of Danish LA21 Politics

Implementing the Earth Summit's declaration chapter on LA21 was, in Denmark, initiated by dedicated politicians at either the national or local levels, most often together with local municipality associations, NGOs and entrepreneurial employees from public institutions and residence areas (Holm 2007). During the mid-1990s, a policy window for value-based local politics under the LA21 umbrella was detected and developed. Front-runner municipalities and counties that previously had been environmentally or socio-culturally active began to show interest in supporting combined social and green practical activism for the re-building of the community outside the global capital markets. This included, among other things, mobilising clients and citizens in making renewable energy utilities, developing energy-saving measures, renewing urban areas with ecology projects, reducing the amount of traffic, etc. These dispersed environmental activities, originating as well in a horizontal process as by state initiatives, served to inspire a number of publicly launched initiatives concerning LA21 projects among the municipalities.

The tradition of a decentralised public administration and the consensus-seeking approach, together with a tradition of 'popular enlightenment' provided a favourable landscape for the implementation of LA21. In the same way, the local government's tradition, characterised by public participation in local planning together with a local environmental policy of integrated pollution-control measures (since 1991), has prepared a socio-technical regime among local authorities for the implementation of Rio's LA21 mandate. In addition, the multi-partisan tradition that incorporates plural interest groups in the design and implementation of local policies together with a comprehensive number of local, green 'do-it-yourself' experiments has made it relatively easier for LA21 officials to initiate LA21 projects with a considerable degree of public interest (for details see Holm 1999). As is clear, these front-runner environmental activities inspired a number of publicly launched initiatives in practice. The initiatives and networks formed an important basis for the subsequent initiation of LA21 projects in the municipalities. Of most importance is that the experiments revealed a new paradigmatic way for developing a separate path in environmental policy, where supporting bottom-up approaches formed new visions for local development and social mobilisation. New partners were found for a number of environmental areas that were not under the rules of environmental acts and regulations. They formed the basis for a change towards including citizens, NGOs and authorities in more comprehensive and constructive efforts to re-build cities and infrastructures. Resource accounting, quality of city life and environmental goods became a positive focus, instead of protecting the environment through restrictions on activities.

Following the Rio declaration, the Danish Government took two years before it initiated the implementation of the Rio mandate on LA21. In the autumn of 1994, the Ministry of the Environment and Energy (MEE) assigned two academics the responsibility of initiating LA21 campaigns among the municipalities and counties. They were placed in the department of National Spatial Planning and have been most active in travelling round the country in cooperation with the municipalities. There have been no efforts to include other state sectors in LA21 activities – not even the departments for industry, agriculture, cleaner products or pesticides.

The MEE has, since the aforementioned initiation, adapted a collaborative regime with relevant interest groups, the National Association of Local Authorities (KL) and the Association of County Councils (Amtsrådsforeningen i Danmark (AF)). A network of active municipalities has functioned as the MEE's access to local authorities when information and other types of campaign have been launched. MEE, KL and AF requested local authorities to start (voluntarily) using five terms of reference as guidelines for their planning processes. In summary, the local authorities were to initiate: (1) cross-sectoral and holistic efforts; (2) active public participation; (3) a cradle-to-grave perspective; (4) global concern; and (5) a long time span perspective. These terms of reference have become a crucial point for many local authorities in the Danish, top-down, implementation process of LA21.

We may describe the following development paths in Danish LA21 policies (Holm 2007) as shown in Table 10.1.

Table 10.1   Danish LA21 developments

1) *1992–1996, Pioneer and generic period*: Rio-conference, kick-starting Aalborg-charter in 1994, networks based co-operation among public authorities, campaigns by using the few LA21-active pioneers. Governmental campaigns for thorough sustainable transition.

2) *1996–2000, Bottom up initiatives and expansion*: Increase in number of LA21-active municipalities; public participation in plans, hearings and a number of environmental projects in housing areas, municipal institutions, transport, catering and transport. A spirit of deliberation and 'let 1000 flowers grow'. Network based experimenting with sector crossing, mobilising initiatives. Green Guides support scheme (Den Grønne fond), state financed nature interpreters and a subvention scheme for green jobs (Den Grønne Jobpulje) provide actors and resources. LA21-officials in satellite-positions bridges citizens groups and municipalities.

3) *2000–2003, Decline, formalising and change of course*: Downplaying LA21-efforts and often re-labelling them as green management and the like, closing of national EPA-funds for LA21-investments; blocking a bottom up process of forming a national Agenda 21, marks a stop for networks based cooperation. An enacted demand for mandatory Agenda 21-strategies in the parliament provides a basic political commitment. Growing differentiation among front and laggard LA21 municipalities.

4) *2003–2009, Private rebuilding efforts and new climate foci*: self generated networks among frontrunner municipalities, private consultancy based networks with supplementary training, division of labour on special focus areas, in-house greening, ecological construction projects, new organising in greater municipalities and regions with beginning far-sighted ends in municipal plan documents. A number of municipality climate-plans are developed by Local Agenda 21 staff including policy areas as energy use in housing, water management.

Although public participation and public involvement are old issues in Denmark, LA21 activities have nevertheless advanced considerably more in integrating laypeople's opinions in local environmental policies than was possible within conventional public participation efforts. From 1996 to about 2002 between 30–50 per cent of the municipalities established citizens' fora and regular public meetings concerning environmental issues, 23 per cent had negotiations with NGOs and about 25 per cent used education for citizens' involvement. Participation concerned in some cases the problem-definition stages, in targets and priorities setting, and some have used participatory processes in the search for visible, socially and locally desirable solutions. But most of the authorities (40–60 per cent) have primarily used LA21 in the expert-lay tradition as a forum for new ways of *informing* their presumably 'ignorant' citizens. Another kind of support from local authorities to public participation is funding citizen initiatives – 20 per cent of the municipalities have used this as a tool (CASA 2004). Public participation has been a paradigmatic turn that maintains: for the moment (2007–2009), Denmark is experimenting with bringing public participation into framing, planning and managing an upcoming number of national parks, and public participation has become part of the role for the more than 300 local nature guides or interpreters. Still, LA21 programmes and activities have only promoted true, shared responsibility and cooperation regimes amongst local citizens and local authorities in the few best cases.

A political modernisation of the vertical and horizontal *organisational structures* has developed and influenced the process. Here the Government and the MEE have been innovative in advancing a horizontal networking strategy for the top-down advancement of LA21 in Denmark. This seems to have benefited new kinds of catalytic approaches, for example, the appointment of the approximately 200 LA21 officials, the approximately 100 green guides and the approximately 300 nature interpreters, not as administrative staff but as semi-public/private entrepreneurs given the responsibility of making experimental innovation, mobilising citizens, revealing our fragile nature, initiating debates and establishing socio-environmental activities. Networking is another case in point. Through contact persons, informal and formal networks, and other collaborative national or regional LA21 activities, LA21 officials and interested citizens have been given the opportunity to exchange ideas and share experiences. There are five centres in 16 municipalities engaged in this kind of committed benchmarking and learning organisations. Since the introduction of the Danish Spatial Act of 2001 it is now mandatory for all municipalities to produce and publish a statement on their LA21 strategy, containing five target areas, among which one concerns public participation that has to be reflected upon. It is a reflexive institutional device as it is up to the municipalities to develop their own approaches to LA21, as the law and the MEE do not demand certain aims, strategies or methods.

The official plan document, *The Local Agenda 21 Strategy Tool*, forms a policy option for drawing up sector- or community-based *transition* arenas, but an evaluation reveals that most often the overall community goal's role is very soft and general, meanings goals with no specified enforcement strategy (CASA

2004). For the sector parts, the published strategies, according to the CASA 2004 evaluation, are mostly concerned with traditional environmental issues such as waste (65 per cent of the strategies), energy (65 per cent), chemicals (50 per cent), biodiversity (50 per cent), and drinking water (50 per cent). When it comes to our theoretically informed question of transition, a paradigmatic turn is, however, traceable in the deliberate greening of public administrative units, of ecological and green public purchasing (43 per cent), and of servicing, as well as in efforts for greening lifestyles (22 per cent). LA21 has been turning many municipalities into partly green enterprises, supporters of green lifestyles and a forum for ensuring citizen participation. New niche systems in renewable energy supply, in new types of energy-saving dwellings, in recycling, in integrated wastewater regulation, etc. have in several cases succeeded, but many isolated projects of urban ecology, etc. have often been too fragile to survive in the long run, exactly because they were not supported by a larger socio-technical regime. The LA21 initiatives have only been 'exterior *add-ons*' to existing local environmental, social and business policies, not addressing cross-sectoral sustainability policies. This may also have been successful for the shaping of new discursive alliances from below, as bureaucratic and corporate interests were not threatened.

Concerning transition, new structural measures for sustainability are not generally apparent within LA21 in Denmark. Thus, even though many small islands of innovations have come into being, there has been a lack of major strides to sustainable structural patterns of manufacturing and consumption. Local politics have, with a few exemptions, not integrated a global dimension, the cradle-to-grave perspective or long time span perspectives. This is an effect of the consensus-seeking and let-a-1,000-flowers-grow policy style of Danish LA21 in general, but also stems from a deliberate avoidance of including industry-, agriculture- and transport-related environmental regulation. But our explorative study of some of the best cases (Holm 2003, Holm, Stauning and Søndergaard 2009) reveals that in many of these municipalities, we have seen how the more conventional political, economic and technological approaches to environmental regulation are being supplemented by more cultural and everyday-life dimensions. Here citizens' moral and cultural rationalities come to play a new and important role. This has facilitated the 'ecologisation' of a considerable number of local projects, not least with respect to social relations. Some of the best cases that have had a high score on LA21 methods and content also happen to be the ones that have experimented more with new ways of public involvement and participation. In other words, we can conclude that pleas for politicians to address the 'real' targets and point to new structural means are not realistic if not embedded in a new public discourse concerning transition, but this discourse will also have to develop outside public institutional arrangements.

Categorising the Danish LA21

Of most importance for this chapter is that the LA21 experiments have revealed a new paradigmatic way for developing a separate path in environmental policy, where political mobilisation forms new *visions for local development, community cohesion and social welfare*. New partners were found for a number of environmental areas (e.g. social care institutions, car drivers, infrastructure managers, dwelling constructors, gardeners in public green areas, etc.) that were not under the rules of earlier environmental Acts and regulations. They formed the basis for a process of change – including citizens, NGOs, business and authorities – in more comprehensive and constructive efforts to re-build cities and infrastructures, to engage health and social care activities in greening. Resource accounting, quality of city life and environmental goods became a positive focus in an optional-oriented policy feature, instead of protecting the environment through restrictions on activities, the 'bads'. Governing LA21 has, beside a lot of window dressing and simulation politics (Blüdhorn 2005), been a vehicle for new types of governance beyond the scope of the *rechtstaat or the welfare state*; LA21 policies in Denmark have fostered imaginative community policies, experimentation, partnership politics, and corporate and civic deliberation especially among front-runner municipalities.

The Danish LA21 politics can be said to be inspired by social ecology (Forsberg 2002) *niche* efforts of:

- creating protected spaces of alternative energy supply and a path development of resource saving;
- recycling streams and infrastructure;
- test beds for urban ecology show cases;
- green consumer and public purchasing movement for a local ecological niche market;
- a basic needs-driven culture to lower materialistic consumption.

This has occurred within a *landscape* of:

- a decentralised welfare economy and government structure;
- high political mobilisation regarding ecological issues;
- entrepreneurship and informal networks among civic servants, social movements and researchers;
- limitations in the ecological modernisation type of industrial environmental regulation (so far);
- willingness to be in front globally (sic!);
- scattered globalised division of labour in food, construction, etc.

The niche efforts may be perceived as change efforts to break with the systemic lock-in of:

- consumption and economic growth-oriented ecological modernisation in environmental policy;
- sector and functional differentiated environmental regulation;

- national and international dispersed environmental infrastructure separated from industry, farming, etc.; and
- generally 'leaving it to the experts' culture.

In a crude schematic overview we have in Danish LA21 policies of the past 10 years, found the following:

Table 10.2   New types of politico-administrative configurations in LA21-based policy

Boundary crossing efforts: Sector policy coordination between local environmental and social/cultural/technical policies; environmental assessment and priority tools in public construction; innovative projects crossing private-public-civic sector borders.

Deliberation in strategic plans and policies: Strategic LA21 development plans for an imagined community future in a sustainable direction; sector and spatial strategy plans.

Mobilisation and framing mindsets: Forming and coaching partnerships; generating public participation in planning, in debates, in managing; discursive campaigns; greening by education.

Greening in-house management: Urban ecology in municipality owned areas; energy and environmental management schemes in institutions; green public purchasing; organic food supplies to public meals, etc.

Responsive networks governance and experimentation among NGO's, local administrators, and corporations by: Imaginative Community devises; inviting networks in local policy formulations; funding and assisting entrepreneurial networks in projects; tools for ecological self-evaluation among institutions; socio-ecological indicators.

Benchmarking governance: Groups of green municipalities committing themselves to continuous improvements, innovation sharing; establishing partnership with corporate green networks; initiating green flag label for voluntary greening efforts among schools, and blue sea label/flag for harbour and coast-line areas.

Framing community coherence: Strengthening green community identity in culture; framing and using local nature and resources; creating social meaningful and ecological jobs in the neighbourhood; supporting local clusters and shortening product chains in an alternative economy.

Altering hierarchical rule games: Horizontal networking among national state agencies, associations of local governments and local municipalities; leaving legal forms of top-down regulation, to the benefit of emerging 'optional' politics.

Participating in international hybrid networks: Local authorities and NGO's have by the Rio process and LA21 commitment formed many new Baltic activities and arrangements on various institutional levels.

We are witnessing a mixture of on the one hand a number of front-running municipalities that are even deepening external cooperation and benchmarking. On the other hand, a growing number of municipalities in which the LA21 effort is either downgraded or integrated into normal public environmental-technical services or local investment programmes. There has been a decrease in the public participatory schemes and efforts, and part of the pioneering spirit of deliberation has vanished in many areas. On the other hand, the municipalities with stable and deepening LA21 activities have committed themselves into community development and experimentation of a new kind, where more sector-integrated approaches flourish which is especially the case on climate mitigating efforts and plans (Bjørnæs et al. 2004, Holm, Stauning and Sønderggaard 2009).

Conclusion

From our studies, we have derived some interesting challenges to conceptualise some of the deficits in the current development of LA21 politics in Denmark. A better-targeted character of niche management to transform socio-technical regimes will have to be performed on a larger scale if not losing many of the promising scattered efforts and insights. New and broadened networks seem necessary to develop, foster and supply a more step-by-step development strategy if we are to overcome the small islands of niche experiments. Thoughts about cultural politics and reflexive modernisation have pointed to the need for a deliberate effort to sponsor the development of new mind frames and citizens' interest in community development. Besides, when more business-oriented types of ecological modernisation take over LA21, the encouragement of new cultural horizons in qualitative broadened comprehension of environment will be missing. The niche strategy has to gain more successes and experience has to be condensed to let research point at promising results and learning for next generation of more strategic LA21 efforts. The challenge is to enhance a meaningful or relevant sustainable cohesion of the community, in a world of disappearing relationships between citizens, manufacturing and authorities.

We might say that the Danish LA21 efforts for a radical institutional change of environmental policy has revealed a number of interesting breakthroughs that on the one side were deemed to run out of steam after the withdrawal of resources and attention. On the other hand, a lot will maintain (horizontal networking, ecological and green purchasing, environmental and energy management, public hearings, visionary and strategic planning) and be dispersed to other policy areas. The ecological transition efforts of greening institutions and everyday life – with experiments of new partnerships, citizen participation, practical experiences in car-sharing, green purchasing, the public provision of organic meals, greening households and institutions, etc. – have thus enhanced new identities, cognitive frames and practical knowledge of what may be done. But to accelerate and accumulate the many interesting experiments of forming values, a deliberative move beyond niche efforts is necessary. This means the forming of transition management in order to move into an orchestrated endeavour of a kind we have only seen some small signs of in organic food supply and purchasing systems, in renewable energy systems, in ecological housing, and in waste and recycling infrastructure. Here it seems that a break with the hitherto logic of appropriateness in local environmental politics and administration must be searched for.

We have witnessed a number of fall-outs and steps backwards in this deliberative regime of local politics for a sustainable culture. No wonder, as the sustainable development policy effort of LA21 is embedded in an aggressive growth economy, a polity system, a political hegemony of growth-based welfare, and a cognitive and social culture, which saliently filters out anything that is not 'realistic'. The dominant impression is that the transformative impacts on core LA21 issues such as environmental sector coordination, transforming consumerism, the scale of societal metabolism with nature, democratic participation, cleaner technology systems and services – are modest (Bjørnæs et al. 2004). The poor LA21 results are often blamed on the lack of political will and resources, window dressing, counter-productive power games on centralisation and decentralisation (Riordan and Voisey 1998, Lafferty 1999, Norland, Bjørnæs and Coenen 2003). But above that we will point at a systemic lock-in and a mind frame to which any kind of deliberative ecological politics would easily have to surrender.

In contrast to many politically involved LA21 observers, I do not think it will be sufficient to look alone for implementation strategies and political will to enforce sustainable development strategies in LA21, or to mobilise a new revival among the staff of LA21 employees and politicians, or to call for a thousand participative future workshops. Rather, political modernisation strategies that reveal ambivalences and dilemmas, new political identities and transition agendas that may be attractive for the inhabitants, institutions and professionals have to be searched for if policy aims regarding sustainability are ever to be achieved in LA21. The manifold experimentations on bottom-up and horizontal network governance, sector coordination, greening lifestyles and institutions of LA21 in Scandinavia have precisely their importance in contributing to a search for new networks and identities in a political modernisation process that may form a beginning of an environmental transition. Here political modernisation takes place by searching for new boundaries and links between state officials, citizens, companies and NGOs in developing alternative service systems and technological systems, and in emerging new patterns of social behaviour. But these local experiments and identity developments have to be developed into managed experiments for the sake of changing consumer habits, mobility and the local economic behaviour of existing dirty industries on a larger scale. The many LA21 experiments will have to be coordinated not only horizontally but also vertically to change the wider structures and regimes of mobility, of construction, of housing, of certain division of labour. It is a task for public managers, politicians, educators, LA21 staff and researchers to uncover and compile the experiences from these efforts in order to condense options and blueprints for LA21 politics in the future. This is also in line a conclusion from a Scandinavian seminar among LA21 practitioners, politicians and researchers in Gothenborg, Sweden in 2004 (Bjørnæs et al. 2004). From that point of view we, here, want to point at the challenge of LA21 as a means for sustainable development in linking up for further transition processes on a larger scale.

References

Agger, A., Hoffmann, B., Jensen J.O., Læssøe, J. and Madsen, K.D. 2005. *Evaluering af Københavns kommunes Agenda 21 centre og sattelitter*, Evalueringsrapport for Københavns Kommune. Copenhagen: Københavns Kommune.

Beck, U. 1994. Self-dissolution and Self-endangerment of Industrial Society: What Does it Mean?, in *Reflexive Modernisation*, edited by U. Beck, A. Giddens and S. Lash. Cambridge: Polity Press.

Bjørnæs, T., Boas, M., Eckerberg, K. and Holm, J. 2004. *Hur står det til med det lokala Agenda-21 Arbetet i Norden?* [Online]. Paper for the ACTION CONFERENCE Jordens Agenda – Nordens Agenda, Göteborg, 2-3 September. Available at: http://www.ieh.se/nordensagenda/dokument/forskningsrapporter. pdf [accessed: 9 November 2009].

Blühdorn, I. 2005. *The Politics of Simulation: Eco-politics in a Post-Environmental Era*. Paper to MISONET Research Seminar, Copenhagen, April.

Brown, H.S., Vergragt, P., Green, K. and Berchicci, L. 2003. Learning for Sustainable Transition through Bounded Socio-technical Experiments in Personal Mobility. *Technology Analysis & Strategic Management*, 15(3), 291–315.

CASA. 2004. *Undersøgelse af det lokale Agenda 21-arbejde, Dansk Status 2004*. Udarbejdet af CASA for Miljøministeriet, Skov og Naturstyrelsen, Landsplanafdelingen, København.

Dahlgren, K. and Eckerberg, K. 2005. *Status för Lokal Agenda 21 – en enkätundersökning 2004*. Umeå: Hållbarhetsrådet.

Forsberg, B. 2002. *Lokal Agenda 21 för hållbar utveckling – en studie av miljöfrågan i tillväxtsamhället*. Umeå: Statsvetewnskabliga Institutionen, Umeå University.

Geels, F.W. 2002. Technological Transition as Evolutionary Reconfiguration Processes: A Multilevel Perspective and Case Study. *Research Policy*, 31, 1257–74.

Geus, M. 1996. The Ecological Restructuring of the State, in *Democracy and Green Political Thought: Sustainability, Rights and Citizenship*, edited by B. Doherty and M. de Geus. London: Routledge.

Grin, J., Rotmans, J. and Schot, J. 2009. *Transitions to Sustainable Development*. Amsterdam: KSI Paper.

Hajer, M.A. 1995. *The Politics of Environmental Discourse: Ecological Modernization and the Policy Process*. Oxford: Oxford University Press.

Hajer, M.A. 1996. Ecological Modernization as Cultural Politics, in *Risk, Environment & Modernity: Towards a New Ecology*, edited by S. Lash, B. Szerszynski and B. Wynne. Oxford: Sage.

Hajer, M. and Fischer, F. 2000. Beyond Global Discourse: The Rediscovery of Culture in Environmental Politics, in *Living with Nature*, edited by M. Hajer and F. Fischer. Oxford: Sage.

Held, D. 1995. *Democracy and the Global Order: From the Modern State to Cosmopolitan Governance.* Oxford: Polity Press.

Holm, J. 1999. LA21 and Political Modernisation: Toward a New Environmental Rationality in Denmark?, in *Implementing LA21 in Europe – New Initiatives for Sustainable Communities*, edited by W. Lafferty. Oslo: EU Commission and ProSus, Oslo University.

Holm, J. 2004. Participation and Consensus-seeking for Sustainable Development?, in *Debating Participation*, edited by R. Gustavo, R. Søren Lund and M. Michael. Copenhagen: Royal Danish Academy of Fine Arts, School of Architecture.

Holm, J. 2007. Eksperimenter i lokalsamfundets miljøomstilling – 12 års Lokal Agenda 21-arbejde i Danmark, in *Økologisk modernisering på dansk – brud og bevægelser i dansk miljøindsats*, edited by Jesper Holm, L.K. Pedersen, J. Læssøe, A. Remmen and C.F.J. Hansen. Copenhagen: Frydenlund.

Holm, J., Hansen, O.E. and Søndergård, B. 2003. Environmental Policy and Environment-oriented Technology Policy in Denmark, in *Environmental and Technology Policy in Europe – Technological Innovation and Policy Integration*, edited by J. Geerten, I. Schrama and S. Sedlacek. Dordrecht: Kluwer.

Holm, J., Stauning, I. and Sønderggaard, B. 2009. *Klimaændringer og innovation i byggeriet.* [Online]. Available at: http://www.klimabyggeri.dk [acceessed: 15 August 2009].

Jacobsen, S. and Bergek, A. 2003. Transforming the Energy Sector: The Evolution of Technological Systems in Renewable Energy Technology. [Online]. Paper to the conference 'Governance for Industrial Transformation', Berlin, 5–6 December. Available at: http://www.fu-berlin.de/ffu/akumwelt/bc2003/download.htm [accessed: 9 November 2009].

Jänicke, M. 2000. *Ecological Modernisation: Innovation and Diffusion of Policy and Technology.* Berlin: FFU working paper.

Jänicke, M. 2006. *The Rio Model of Environmental Governance: A General Evaluation.* Berlin: FFU-report.

Joas, M. 2001. *Reflexive Modernisation of the Environmental Administration in Finland.* Åbo: Åbo Akademi University Press.

Joas, M. 2001. Milöpolitik och ekologisk modernisering i Finland, in *Från teknik til etik i Miljöförståelse i universitetspolitik*, edited by E.M. Blomqvist. Åbo: Åbo Akademis Forlag.

Jørgensen, U. and Lauridsen, J. 2007. Kan miljøpolitik stimulere dynamisk teknologiudvikling?, in *Økologisk modernisering på dansk*, edited by J.

Holm, L.K. Pedersen, J. Læssøe, A. Remmen and C.F.J. Hansen. Copenhagen: Frydenlund.

Kemp, R. and Loorbach, D. 2003. *Governance for Sustainability Through Transition Management.* Paper to the conference Governance for Industrial Transformation, Berlin, 5–6. December. [Online]. Available at: http://www.fu-berlin.de/ffu/akumwelt/bc2003/download.htm [accessed: 9 November 2009].

Kemp, R. and Rotmans, J. 2001. *The Management of the Co-evolution of Technical, Environmental and Social Systems.* Paper to the conference Towards Environmental Innovation Systems, Garmisch-Partenkirchen, 27–29 September.

Kemp, R., Schot, J. and Hoogma, R. 1998. Regime Shifts to Sustainability Through Processes of Niche Formation: The Approach of Strategic Niche Management. *Technology Analysis & Strategic Management*, 10(2), 409–29.

Kemp, R., Weber, M. and Schot, J. 2001. *Transitions to Sustainability through System Innovations.* Paper to International Expert Meeting, 2001, Eindhoven: Twente University.

Læssøe, J. 2000. Folkelig deltagelse i bæredygtig udvikling, in Dansk naturpolitik i bæredygtighedens perspektiv. København: Naturrådets temarapport nr. 2.

Lafferty, W.M. 1999. Status of LA21 in Europe, in *Implementing LA21 in Europe*, edited by W.M. Lafferty. Oslo: ProSus, Oslo University.

Lafferty, W.M. 2001. Conclusion, in *Sustainable Communities in Europe*, edited by W.M. Lafferty. London: Earthscan.

Mortensen, J.P. 1999. *Integration af kredsløbstankegang og forbedringspotentialer i miljøgodkendelsessystemet.* Ph.d. dissertation. Roskilde: Roskilde University.

Murphy, J. and Gouldson, A. 2000. Environmental Policy and Industrial Innovation: Integrating Environment and Economy Through Ecological Modernisation. *Geoforum*, 31(1), 33–44.

Murphy, J. 2000. Editorial – Ecological Modernisation. *Geoforum*, 31(1), 1–8.

Norland, I.T., Bjørnæs, T. and Coenen, F. 2003. *Local Agenda 21 in the Nordic Countries – National Strategies and Local Status.* Report No. 1/03. Oslo: ProSus, University of Oslo.

Quist, J. and Vergragt, P. 2003. Backcasting for Industrial Transformations and System Innovations towards Sustainability: Is it Useful for Governance. Paper to the conference Governance for Industrial Transformation, Berlin, 5–6 December. [Online]. Available at: http://www.fu-berlin.de/ffu/akumwelt/bc2003/download.htm [accessed: 9 November 2009].

Rennings, K., Kemp, R., Bartolomeo, M., Hemmelskamp, J. and Hitchens, D. 2003. Blueprints for an Integration of Science, Technology and Environmental Policy. [Online], in *(Blueprint), the Improving Human Potential Programme: The Strategic Analysis of Specific Political Issues (Strata).* Available at: http://www.blueprint-network.net [accessed: 9 November 2009].

Riordan, T. and Voisey, H. 1998. *Sustainable Development in Western Europe: Coming to Terms with Agenda 21.* London: Routledge.

Rip, A. and Kemp, R. 1998. Technological Change, in *Human Choice and Climate Change, vol. 2*, edited by S. Rayner and E.L. Malone. Columbus, OH: Batelle Press.

Rotmans, J. 2001. *Transitions and 'Transitions Management': The Case for a Low Energy Supply.* [Online]. Available at: http://www.icis.unimaas.nl/publ/downs/01_13.pdf [accessed: 9 November 2009].

Rotmans, J. 2003. *The Dutch Knowledge Network on System Innovations and Transition.* Paper to the conference Governance for Industrial Transformation, Berlin, 5–6 December. [Online]. Available at: http://www.fu-berlin.de/ffu/akumwelt/bc2003/download.htm [accessed: 9 November 2009].

Smith, A. 2003. Transforming Technological Regimes for Sustainable Development: A Role for Alternative Technology Niches. *Science and Public Policy*, 30(2), 127–35.

Smith, A., Berghout, F. and Stirling, A. 2003. *Transforming Industrial Sustainability Under Different Transition Contexts.* Paper to conference of Governance for Industrial Transformation, Berlin, 5–6. December. [Online]. Available at: http://www.fu-berlin.de/ffu/akumwelt/bc2003/download.htm [accessed: 9 November 2009].

Søndergård, B., Holm, J. and Hansen, O.E. 2004. Ecological Modernisation and Institutional Transformation in the Danish Textile Industry. *Journal of Cleaner Production*, 12(4), 337–52.

Truffer, B., Metzner, A. and Hoogma, R. 2002. The Coupling of Viewing and Doing: Strategic Niche Management and the Electrification of Individual Transport. *Greener Management International*, 37, 111–24.

Vellinga, P. 2003. *Transition Research to Address Global Environmental Change.* Paper to the conference Governance for Industrial Transformation, Berlin, 5–6 December. [Online]. Available at: http://www.fu-berlin.de/ffu/akumwelt/bc2003/download.htm [accessed: 9 November 2009].

Weale, A. 1992. *The New Politics of Pollution: Issues in Environmental Politics.* Manchester: Manchester University Press.

Weaver, P., Jansen, L., Grootveld, G.V., Spiegel, E.V. and Vergragt, P. 2000. *Sustainable Technology Development.* Sheffield: Greenleaf Publishing.

United Nations. 1993. *Earth Summit: The United Nations Conference on Environment and Development (UNCED).* London: Graham & Trotman.

Chapter 11

Democracy and Sustainability: A Lesson Learned From Modern Nature Conservation

Laura Tolnov Clausen, Hans Peter Hansen and Esben Tind

Introduction

Sustainability has played an increasing role in the efforts of international society to deal with global environmental problems. Sustainability was emphasised by the report of the Brundtland Commission, 'Our common future' (Brundtland 1987), and the term became a keyword on the international political agenda of the UN's conference on environment and development, the so-called Earth Summit held in Rio de Janeiro in 1992. When deconstructed, one can say that the term has become a critical political category on the one hand, and a regulative strategic tool within public planning on the other. It is implicitly critical in political terms because of its questioning of the aggregated consequences of utility maximisation in terms of the destruction and exploitation of natural resources, and in terms of social and economic inequalities, and poverty. The regulative strategic dimension of sustainability concerns its orientation which is supposed to contribute to better and more holistic regulation, not only dealing with the natural aggregated consequences of human behaviour but also with the driving forces behind such human behaviour. Since the beginning of the 1990s, sustainability has been a central feature of several international conventions and agreements. The Convention on Biological Diversity is but one example of how sustainability has become a keyword within nature conservation policy:

> ...the sustainable use of its components and the fair and equitable sharing of the benefits arising out of the utilization of genetic resources, including by appropriate access to genetic resources and by appropriate transfer of relevant technologies, taking into account all rights over those resources and to technologies, and by appropriate funding. (UNEP 2000)

In the Convention, sustainability is often used to describe the strategic focus of the Convention. The introduction of the concept has challenged the more classical approach to nature conservation by emphasising that the protection of our natural resources must be achieved by a large-scale change in how we use the resources.

Aspects other than biological ones have to be taken into consideration – factors such as production, the use of technology, the economy, equality and justice all have to be integrated into a conceptualisation of biodiversity. Politicians and planners are forced to search for new planning strategies due to the broad public and political recognition that we as modern societies have not so far been able to deal properly with environmental problems. In view of this recognised inadequacy and the pressure from the public, 'participation' has been central in the search for new strategies within the last 15–20 years. It was highlighted at the Earth Summit and explicated in the 'Local Agenda 21' (LA21) programme. From then on, the concept of participation has been combined with sustainability and played a key role around the globe. However, when we examine contemporary policies and public planning related to environmental issues, it seems that politicians and public planners in all parts of the world have major difficulties integrating participation and sustainability within public planning. In this chapter, we will argue that a new and more profound democratic understanding of sustainability is needed to provide the necessary changes in society. This understanding includes a utopian dimension which contrasts with a more strategic and risk-oriented approach. To illustrate our point we will use a case taken from the field of nature conservation, namely the recent development of the concept of national parks in Denmark. This is used to demonstrate some of the structural issues at stake when dealing with participation and sustainability within public planning.

Our chapter is divided into five sections. In the following two sections – 'Sustainability from a Critical and Strategic Point of View' and 'Sustainability as a Democratic Term' – we will deconstruct the term sustainability from both critical and strategic points of view, and demonstrate that the two different dimensions of the term sustainability will create conflicting meanings if not thoroughly considered when creating policy reflected. In sections three and four – 'National Parks in Denmark' and 'Democratic Sustainability and National Parks' – we will describe the process of establishing national parks in Denmark and the structural problems of this process. In section five, the 'Discussion', we will use the case study to discuss the need for profound changes in our approach to sustainability.

Sustainability from a Critical and Strategic Point of View

The Brundtland Report highlights many problems related to the way we produce and distribute resources and, importantly, those tied to complex ethical questions like those surrounding the political, social and cultural inequalities that exist among people. In so doing, the word sustainability becomes a normative term which relates to ethical questions that take their starting point in issues concerning freedom, equality and justice. But the use of the word sustainability implies more than just desirable visions; it also contains a critique of the way we plan development today:

> A plan or political action, national or international, that does not include
> sustainable development as its goal, is hard to find. (Elling 2008: 17)

The value oriented and critical dimension of sustainability, which is approved by the international community, raises a fundamental question: How can or should ethical and moral philosophical issues be transformed into action? Is it at all possible to integrate all questions in all their complexity and endless possibilities of interpretation into public environmental planning and regulation? Whilst 'ecological modernisation' (e.g. Mol 1996, Hajer 1995, Elling 2008) has attempted this, the question which still remains to be answered is whether it is possible to integrate the complexity of the world into an environmental paradigm?

When deconstructing the definition of sustainability we realise that *time* is a key element of the term, and that *needs* and *future*, become 'principals of negotiation', and thereby meanings that are left for the specific social and cultural contexts to define. The 1992 UN programme, LA21, however, attempts to give sustainability a legal status. The 'principal of negotiation' – 'needs' and 'future' – are replaced by what the Danish philosopher Gorm Harste calls a 'constituting of causality' (Harste 2000) suggesting that sustainability can be read in 'Mother Nature', and 'indicators' of nature and the environment become determinants in the interpretation of the meaning of sustainability. This gives natural science a superior status and therefore becomes a part of the strategic dimension of sustainability. According to Harste, the idea that 'Mother Nature' has the answer to sustainability creates two fundamental problems. (1) Political action has to be oriented forward in time, but modern scientific knowledge (about nature) is per definition oriented backwards in time. Thus, if political actions are based on indicators which are built on scientific knowledge, action will by definition be late. (2) This idea of 'Mother Nature' giving us the definition of sustainability creates a problem of integrating one system's (natural science) way to interpret sustainability into other systems.[1] Such integration is impossible due to the fact that social, cultural, scientific and political systems (Luhmann 1982) are self-referring; each operating with different types of rationality and time perspectives. When interpreted by different systemic rationalities, sustainability will automatically be reduced to 'reflective interpretations of futures which should not take place'. Instead of being a starting point for the thinking about alternative sustainable futures, sustainability will be about 'avoiding crises'. Thus the *utopian element* understood as the potential ability to formulate common visions for a better future will suffocate and sustainability will become a dystopian term with an implicit avoidance of certain futures. In this perspective, the strategic use of sustainability will become a positive category of risks and therefore not provide a solution to the causes of the problems.

The same critique of the implementation of sustainability has been made by Wolfgang Sachs who argues that sustainability understood as 'sustainable

---

1   In principle, systems have to be understood broadly, including formal and non-formal systems, public as well as private.

development' no longer represents a hope for the future and therefore is meaningless (Sachs 2000). Sachs puts his point provocatively by concluding that sustainability connected to development is about protecting economic growth against its own risks, instead of finding alternatives to the way we organise our society politically, socially, culturally and so forth. According to Sachs, this development is greater due to the displacement of political power from the nation states to trans-national markets. Elling formulates the same point slightly differently:

> ... and this becomes clearer day by day, is the fact that globalization is dissolving these structures and creating a situation, where total individualization prevails, allowing subservience of everybody to the dictates and demands of the market forces. (Elling 2008: 59)

Altogether one can say that this critical perspective on sustainability highlights a very crucial question: How can public institutions address the ethical and utopian aspects of the term if not through integrating them in the same concept?

Sustainability as a Democratic Term

Gorm Harste suggests an answer to the above question by implying that the ethical and moral dimensions of sustainability require it to be understood as a democratic project for two reasons. Firstly, by giving the natural scientific interpretation of sustainability supremacy we compromise the fundamental role of democracy which is supposed to deal with the complexity of moral, ethical and aesthetic questions on a societal level, and secondly, as mentioned already, there is a strong utopian dimension to the term sustainability. This utopian dimension is in *this* chapter understood as a potential *common* veneration for nature that, given the right circumstances, goes beyond mutual infighting and has the possibility to express horizons of change. According to this understanding, the democratic sphere is the only legitimate political sphere to deal with common visions and utopias for nature and society. In other words, sustainability is per definition a collective challenge, based on differentiation, not integration. It is a challenge that necessitates the experience of nature as something that people have in common in spite of differences among people. And it is a challenge that necessitates dialogue. Several theorists have emphasised democracy and differentiation as the only legitimate way of talking about democratisation of democracy though dialogue (Elling 2008). The sociologist Anthony Giddens is one of them. However, except from referring to the importance of subcultures of society, Giddens does not point towards fundamental changes of praxis. Habermas in contrast points his finger at the main problem: the rationality of the System colonising the Life World (Habermas 1984). For Habermas, colonising (the integration of a certain rationality) is the main problem. His contribution to the theory of the liberal democracy, including his emphasis of the public sphere as the framework for democratic dialogue (Habermas 1976) indicates a possible solution. For Habermas, democratic dialogue is a central part of the public sphere and thereby more than just a subcultural definition. It is the democratic deliberative dialogue which will create the legitimate linkage between the rationalities of systems – economic systems, public institutions and agencies, etc. – and the values of the citizens in their everyday lives. In that perspective, the political definition of sustainability has to be based on a more deliberative democratic approach. Only through a deliberative democratic approach will it be possible to open up the utopian dimensions of sustainability rather than integrative and dystrophic predictions of the future.

In the following section we will describe a case illustrating the need for more profound changes in the way systems define the political agenda on sustainability and participation. The case is built on our research related to the development of national parks in Denmark. This case is interesting within the framework of this book due to the explicit and implicit role of sustainability and participation. The authors of this chapter have been studying and analysing the national park process from its beginning in 2002 until the designation of the first parks in 2007. They have used participatory research methods and qualitative interviews with different actors in the process. The case discussed here illustrates some of the contradictions within modern public planning when it comes to the terms, participation, democracy and sustainability. What makes the Danish case story interesting? In most European and global processes of establishing national parks it is possible to recognise a strong centralistic or expert orientation; somehow national parks have been considered as being a way of protecting nature from local exploitation. Consequently, the social and democratic aspects of national parks have been reduced to the degree of participation which is necessary to avoid too much conflict with the local citizens and to making the implementation of the national park regulations more manageable. In the Danish case, the point of departure was, however, somewhat different. The process was begun with the publication of the so called 'white paper' which meant that the system did not from the very beginning demand any specific solutions and consequently the Danish case opened up the possibility for a potential democratic experiment. That is why the Danish case is unique and globally interesting.

National Parks in Denmark

As a result of the recommendation by the Organization for Economic Cooperation and Development (OECD) (OECD 2000), a political committee established by the Danish Government in 2000,[2] the so-called Wilhjelm Committee, took up the question of establishing national parks as a tool for improving biological diversity in Denmark. In 2001, the Committee agreed to suggest seven areas that potentially

---

2   The Wilhjelm Committee was established by Svend Auken (Social Democrat Party), Minister of the Environment at the time.

could become national parks. Exactly how the suggested areas, of which several were already highly protected and owned by the state, should contribute to the conservation of biological diversity in Denmark was not completely clear, but a democratic and participative perspective was emphasised from the beginning. In 2002, the Danish Government decided to start a programme of pilot projects prior to the final decision on national parks in Denmark. The purpose of the programme was to give local communities two years to, firstly, decide if they were interested in the establishment of a national park in their community and, secondly, to develop some visions of how a park could be integrated with the local community:

> The purpose of the pilot project is, on basis of investigations and local debate, to work out a vision with a plan for the demarcation and the content of a national park on Moen, which can be a part of the foundation for the considerations and decisions regarding the establishments of national parks in Denmark. (...) The pilot project will be carried through on the local level in close cooperation with the local authorities and the Danish Forest and Nature Agency (represented by the State Forest District) under continuous involvement of the local population. (Letter of approval, Minister of Environment 2001, our translation)

Earlier experiments with citizen participation in Danish nature conservation have not been particularly successful, but have instead escalated conflicts between locals and governmental agencies. Within the pilot projects, the Minister tried to ensure a certain degree of local support and understanding of the idea of establishing national parks. To stress the importance of local influence, the Minister of Environment at the time, Hans Christan Schmidt of the Liberal Party, announced that the final political selection of actual park areas would depend on whether the process had been truly democratic and whether a final local agreement was attained. Each pilot project was organised under a local political steering committee, representing different interest groups and political representatives. The specific task of the administration on a daily basis was delegated to local secretariats managed by the regional State Forest Districts and the Chief Forest Officer. The local steering committees were given until July 2005[3] to work out their recommendation to the Minister of Environment.

The Minister of Environment established a 'National Steering Committee' (NSC) representing more than 30 different interest groups and representatives of governmental agencies, which would oversee the different pilot projects. The NSC was given the task of monitoring each pilot project and providing the necessary exchange of knowledge. In some of the pilot projects as well as in the NSC, the job of getting an overview of the results of more than two years of work undertaken by the different pilot projects was given to consultants. On the basis of the NSC's recommendation, the Minister of Environment presented an Act on national parks to the Danish Parliament in spring 2007 (see a simplified illustration of political decision-making process in Figure 11.1).

---

3   Some of the pilot projects were given a three-month extension.

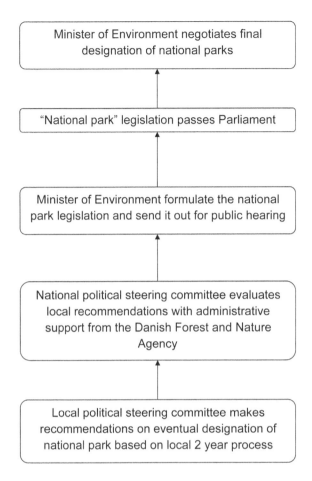

Figure 11.1 Diagram of pilot projects

Note: The diagram illustrates how the two year pilot project process was structured on the local level.

Each pilot project had within its budget the opportunity to design its own course of events including conducting the investigations. Due to the close relationship between the different State Forest Districts and due to the requirement set down by the Minister, the structural backbone of the pilot projects was the same. All the pilot projects used traditional participative methods such as public hearings and surveys in combination with the writing of a large number of expert reports. However, in some of the pilot projects more non-traditional methods have been used, such as future workshops and different kinds of consensus building methods.

In the following, we will describe the progress of one of the Danish national park pilot projects, the project on Moen, an island geographically situated about

100 km (60 miles) south of Copenhagen. Our description is based on systematic documentation of the process, including working papers, minutes from meetings, events and interviews.

Phase 1: The Initial Phase of the Pilot Project

At the local level, politicians, the tourist industry and nature conservation organisations on Moen were thrilled by the idea of becoming a national park pilot project. Immediately after the Wilhjelm Committee suggested Moen as a potential national park, a network of people who could work out a pilot project proposal for Moen was established. In January 2003, seven main normative objectives regarding the pilot project were formulated.

Main Objectives:

1. Secure and expand the beautiful and existent natural and cultural values of north and east Moen.
2. Work for a sustainable use of Moen's natural recourses.
3. Work for the understanding and use of the local qualities of the public, including recreational activities.
4. Work for the sustainable development (social and economic) of the local society.
5. Create a foundation for the formal establishment of Moen national park.
6. Create local ownership to the idea of the national park and demonstrate the potentials and changes/improvements of the idea.
7. Provide input to the national process of establishing a number of national parks.

Apart from a description of the political organisation of the project, the project description was vague about how to create the desired ownership of the project. However, during the preparation of the project description, three public hearings were held. There was a large attendance at all three hearings at which politicians and officials emphasised that everyone who had an interest in being a part of the project could participate in one of the four thematic working groups: *nature and geology; culture and rural development; farming and forestry; and outdoor recreation and tourism.* A relatively large number of citizens signed up for these groups. The main task of the groups was to initiate surveys to develop some collective vision for the park's future according to the main objectives that had been listed. The final output of each working group was to be presented to the political steering committee in a written report (see schematic outline of pilot project process in Figure 11.2).

Even though the idea of the working group was to create a democratic process, the working groups were worried that they could not fulfil this particular task satisfactorily. Therefore it was decided that in addition to the thematic process of the pilot project, a non-thematic process had to be initiated as well. This was done

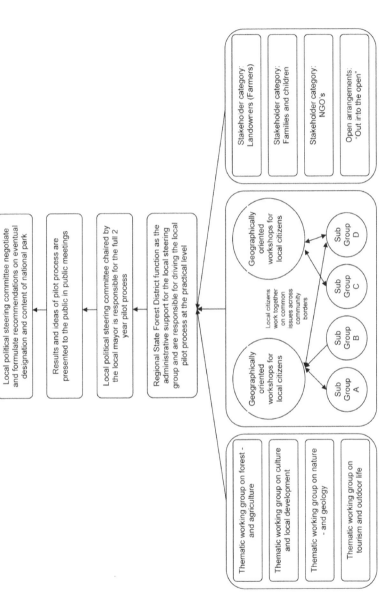

Figure 11.2 Diagram of political decision making in national parks

Note: This diagram illustrates the political decision-making process on national parks from the local to the national level.

by inviting local citizens in a number of geographical areas to participate in future workshops.[4] As a participative method, the future workshop has been developed with special attention being paid to involving all kinds of people regardless of their social and cultural backgrounds and their ability to express themselves verbally. During the workshop, participants try to work out a utopia, to draw an exaggerated picture of future possibilities. Free from inherent necessities and with the use of brainstorming techniques and creative games, the aim is to find and reflect common utopian solutions to identified problems, the basic criterion for these solutions being that they should differ from usual, only rationally (system) orientated problem solutions (e.g. Jungk and Müllert 1987).

During the initial phase of the pilot project, the political steering committee took form and the first meeting of the steering committee was held. It became rapidly obvious that the local farmers were against the project. The position of the farmers led to tensions within the steering committee and external mediators were brought in to depoliticise and mediate the process. To some extent, a better collective understanding of the project among the different actors and stakeholders was created due to the work of the external mediators, but the political steering committee did not have the necessary competence to follow up on the positive results achieved by mediators and create a more positive atmosphere in the meetings that followed and the whole process became politically tense again.

Phase 2: The Working Phase of the Pilot Project

The problems observed in the starting phase continued to haunt the pilot project. The chairs of the thematic working groups worked on a purely voluntary basis and felt a lack of ability and resources to fulfil the task they were given. At the same time, they faced a high drop-out rate from the working groups and a hollowing of the democratic expectations of the pilot project. According to some of the deserting group members, a major reason for the high drop-out rate was the instrumental approach to the work, leaving little space for bringing in other issues of interest, e.g. the more value-oriented issues listed as main objectives. Those laypersons who had signed up because they had veneration for the community including the surrounding nature, suddenly felt marginalised by those people with a more technical approach to the task given. Another problem was the negative influence of the political conflicts with the farmers. Even though some of the farmers were interested in participating in a constructive way, the social control in the farming community made it difficult to prevent the conflict from having a negative impact on the working groups. Much of the debate in the working group was related to this particular conflict. The working groups did produce some suggestions of

---

4 The future workshop was developed by the now late Austrian writer and journalist Robert Jungk after the Second World War. From Robert Jungk, the method has been systematically developed by scholars, among those the two action researchers from the University of Roskilde, Kurt Aagaard Nielsen and Birger Steen Nielsen.

a more value-oriented character. Most significant was the idea – supported by almost all participants involved in the process – that the whole island of Moen should be a part of the pilot project. Despite full support, the idea was hushed up by the political steering committee.

Within the framework of the non-thematic approach, a number of geographically spread future workshops were planned. Two of these were successfully carried through in respect to visions and more utopian ideas. However, the idea of continuing the results of these workshops in a number of specific projects failed, partly because of lack of facilities and partly because of lack of local involvement. The lack of local involvement was also the reason why several other workshops were cancelled. It also seemed as though the thematic and non-thematic processes were in competition with each other. The problem of local involvement in the non-thematic workshops led to the hiring of consultants who had experience in working with laypeople in different kinds of action research projects and different kinds of consultancy tasks. They revitalised some of the ideas that participants from the future workshop already held. Not being able to carry out the initial plan for the implementation of all of the workshops, it was instead planned to support some of the activities started at grass-roots level. Here the focus was on the population of the small island of Nyord (considered a part of the potential national park area) where opposition to the whole idea of a national park was strong.

After having experienced a couple of future workshops, the representatives from the island of Nyord invited the consultants from the University of Roskilde to facilitate some future workshops. Those workshops developed ideas and visions for the future and were presented in a report written by the islanders themselves entitled: 'Nyord as a part of a national park – is that a good idea?' (Foreningen Nyord 2004). In the report the people of Nyord expressed their fear of the national park, but also some of their more visionary perspectives for nature, exciting culture and socio-economic development on the island. On a very concrete level, they suggested a plan of bringing more of the regulation and responsibilities of nature management down to the locals themselves. Instead of delivering the report to the political steering committee, the islanders arranged their own hearing on the island, inviting authorities and politicians to answer some specific and direct questions regarding the political process and the political willingness to work together with locals on these issues.

Phase 3: The Final Phase of the Pilot Project

During the late autumn of 2004, the political steering committee and the pilot project secretariat focused on writing the final report to the Minister of Environment. Due to the problem with the farmers, the working groups and the secretariat held a couple of workshops to find at least some common ground for a collective recommendation of a national park at Moen. The effort did not succeed as other conflicts regarding hunting rights and lack of structural support for the development of small businesses on the island occupied the final phase of the pilot project.

Based on the consultant's initiative, two workshops for all involved in the process were held in September and October 2004. Even though the workshops were formally supported by the political steering committee, political support from the steering committee and from three of the four working groups was limited. The workshops managed to gather some of the local grass-roots organisations. One result of these workshops was the establishment of a local coordination group which was given the task of continuing the work with the visions and ideas developed during the project, regardless of whether a national park was established or not. This coordination group did not get support from the authorities or established NGOs which instead established their own coordination group made up of representatives from the different public institutions, from farmers and a few other NGOs.

On 17 April 2005, all the groups involved – thematic as well as non-thematic – were invited to present their work and visions at a public gathering. In addition, the steering committee insisted on having a traditional public hearing with formal introductory talks by different representatives of the political steering committee. The ensuing debate became a relatively aggressive argument between the local farmers and the representatives for the thematic working groups, most other citizens left the hearing. In the month following the public hearing, the political steering committee and the secretariat finished the final report of the pilot project process at the local level. The chair of the local political steering committee did not succeed in getting all members of the group to agree on a mutually acceptable vision for a national park on Moen. The final report to the NSC and the Minister of Environment contained several minority statements. In spite of the minority statements, the report from the Moen pilot project stressed the importance of future democratic and local regulation of nature management.

Phase 4: Post Pilot Project

After all pilot projects had delivered their final reports to the NSC, the task of summarising the content of all seven reports was handed over to a company of consultants. On the basis of internal negotiations, the NSC made their own report of recommendations to the Minister of Environment, not suggesting any particular areas to be national parks, but suggesting the content of the following framework law. Exactly what content from the reports of the pilot project to the Minister of Environment should be delivered was indistinguishable. The strongest desire expressed by all seven pilot projects, namely that the local communities in the future should have a more direct influence on the regulation and management of nature, did not gain strong support from the NSC and was minimally reflected in the recommendations to the Minister.

After months of political negotiations, the parliamentary system passed the first framework law on national parks in Denmark in the Parliament on 24 May 2007. At the end of 2007 and the beginning of 2008, new political negotiations between parties in the Parliament and the third Minister of Environment, Troels Lund Poulsen from the Liberal Party, resulted in the designation of the first five national parks in Denmark. All the designated parks mainly consist of already highly protected (NATURE 2000) and state-owned areas, and thus the potential improvement of the biological biodiversity in the respective areas appears to be quite limited. From the beginning, most of the designated areas were introduced to maintain and preserve the actual state of the nature and not to create any major changes.

Democratic Sustainability and National Parks

What makes the case interesting in relation to this chapter is its explicit combination of participative democracy, sustainability and political ambition, in a specific political context. Participative democracy was an explicitly political issue which was high on the agenda and was the reason why the Minister initiated seven pilot projects to begin with. Sustainability was explicated in terms of participation, but the whole idea of initiating a national park process had a clear reference to the Convention of Biodiversity and was thereby rooted in the international and more risk-oriented agenda on sustainability. As the case illustrates, there was an explicit political awareness of the necessity to consider social, cultural and economic aspects when establishing national parks in Denmark, which in their essence are about sustainability. From a theoretical perspective one can say that the political ambition of the national park process was to enhance democratic perspectives of sustainability and to create a kind of democratic understanding of sustainability. In the following, we will examine to what extent that was accomplished.

Retrospectively it is clear that the political steering committee on Moen struggled to combine the democratic expectations and the expectations to define a vision for more sustainable development on Moen. One reason for that struggle is that the purpose and the political framework regarding the pilot project were unclear and undefined, and the representatives in the steering committee and the involved authorities had different expectations concerning the task. Another is that politicians, interest groups and officials had different approaches to the creation of a democratic and truly participative process. Regarding the purpose of the pilot projects, the local politicians expected that the establishment of national parks should contribute to economic growth, first and foremost by the strengthening of the tourist industry. The focus on economic perspectives of development contributed to a low priority for changing the way natural resources are used and managed today.

The relatively low priority assigned to nature caused concerns among NGOs and the authorities. From the beginning, the Minister of Environment had emphasised the necessity of local support for the pilot project, but in spite of that, the Danish Forest and Nature Agency as well as the local steering committee had difficulty 'translating' the political demand of participation into the process of the pilot project. Many resources have been spent on thinking 'methods' without the clarification of the role of the citizens and how to distribute power (e.g. Arnstein 1969). Many different stakeholder methods, mediation methods, deliberative methods, etc. were brought into the process. Although the pilot project only minimally succeeded in creating genuine local credibility and awareness about the project, it does not mean local citizens did not contribute to the process, but it meant that only a few of the locals on Moen were involved, and that those actually involved had a hard time being heard.

The involvement of local citizens in thematic working groups was an attempt to engage them in formulating some visions for the future. But instead, the demands for technical surveys – which had a controlling influence on the process – excluded laypeople and made it difficult to develop visions at all. Only minor openings to locally defined natural values were created. In combination with the traditional hierarchical and thematic structure of the working groups, the untrained chairs, the short time available and the ongoing conflict with the farmers meant that the process in the groups suffered quite a lot.

The intent of the non-thematic approach was to create public arenas for local residents, giving them a chance to formulate their own agenda and visions. The whole problem of recruiting locals to these public arenas demonstrated that public participation is not just about changing methods. Within established and relatively well functioning local networks, the non-thematic approach was quite successful. In particular, the citizens on the island of Nyord were inspired to get involved and had a significant impact on the process. The role of the citizens of Nyord demonstrates that it is possible for negative attitudes to be critical but constructive through a non-thematic and utopian-oriented process. In spite of the fact that the citizens of Nyord had a significant impact on the process, this impact was not reflected in the final outcome of the process, basically because the representatives from Nyord did not succeed in transferring their apparently strong position in the middle of the process into the final and highly politicised strategic negotiations.

Discussion

As mentioned in the beginning of the chapter, the concept of 'sustainability' has a critical as well as a strategic dimension. The critical dimension is embedded in the implicit critique of the aggregated consequences of the over-exploitation and the destruction of ecosystems, and in terms of social and economic inequalities and poverty. The strategic dimension is constituted by the regulative action needed to prevent the aggregated consequences of over-exploitation and destruction of natural and environmental resources by imposing a more holistic regulation. It is clear that the introduction of the term sustainability in a theoretical sense has challenged the more classical approach to nature conservation. The concept emphasises that more than biological elements have to be taken into consideration when it comes to the protection of ecosystems and natural resources. Factors such as production, the use of technology, the economy, regulation and the environment are all elements in sustainable public planning. This integrative regulation strategy places new demands on authorities within the field of nature conservation.

In this chapter we have explored how public institutions and agencies deal with these demands in practice. With reference to Gorm Harste, we argued that the integration of sustainability – in its strategic meaning – in different systems is neither possible nor desirable. By integrating sustainability into different and always purposive systems, sustainability will become an undemocratic and dystopian term because these systems deal with the avoidance of certain futures by means of specific and always limited academic and technical skills. In such a regime, participation will be reduced to conflict management or governance and without any utopian impulse in the creation of the future. We argued that sustainability in its critical meaning is not about integration into systems, but about the creation of collective visions for the future. Sustainability is first and foremost about our own and our descendants' future and cannot, within a liberal democratic society, be left to arbitrarily narrow purposive systems and competences to define. Experts and scientists can help to teach us what we do wrong, but they cannot tell us what our future should be. That is a political question, and the reason why sustainability should be understood as a *democratic* term oriented towards the formulation of common visions for a better *future* according to *needs* we human beings define.

The empirical point of departure for this chapter has been the present development of national parks in Denmark – more specifically, the case of pilot project Moen. According to the political rhetorical emphasis on the need for local support for the public regulation of nature and the emphasis on dialogue and participation, the current situation in Denmark could be seen, with reference to Jürgen Habermas, as an attempt to improve communication between the *System* and the *Life World.* Regarding the creation of visions, the pilot project formulated some ideas for the future. However, these ideas and vision have not played any significant role in the political outcome of the process. In that respect, the overall success of the process has been limited when it comes to the creation of a new deliberative democratic agenda on sustainability within the field of nature conservation.

As the case has illustrated, the process did not overcome the traditional technical rationality of expert systems, nor did it fulfil the political ambition of bringing more value-oriented issues into the process – issues which were set out explicitly in the main objectives formulated. The traditional hierarchal structure of the organisation of the pilot project was not able to address the more utopian ideas and visions coming from the non-thematic work in the final report. Despite its ability to incorporate a more utopian orientation during the application of future workshop methodology, the related view on sustainability was not incorporated into the final recommendations of the process. Openings towards new value-oriented approaches were not addressed by the local steering committee. Several reasons can be identified. First and foremost, the process and the participants were caught by the reproduction of classical conflicts related to land ownership. The conventional neo-corporative stakeholder approach endorsed the traditional power structures of the state, agricultural sector, commercial tourist interests and others major NGOs, thereby intimidating more ordinary citizens. One can say that the power given (or taken) by the 'stakeholders' and special interests took the focus away from the 'citizens' and the 'common' or the nature and community as a whole. Supplemented by a thematic and non-integrative approach this had the effect that the working groups were insufficient concerning both social and ecological aspects. Another problem was that scientific instrumentality dominated the thematic working groups and had a marginalising and excluding effect on the participants and resulted in drop-outs. Local citizens felt embarrassed because of their lack of expertise in the process and because of the scientific character of the thematic workshops. Even though some of the remaining participants in the working group were pleased with the output (final report and survey reports) some also felt that the final report was merely a reproduction of the viewpoints and wishes of the chairman and secretary. What in the first place seemed to be a new public participative democratic definition of sustainability, turned out to be the traditional strategic interpretation, to a large extent reproducing the technical and rationalistic views of the governmental agencies and the attempt to create 'futures'. This was to a large extent experienced by the locals as a strategy aimed at creating limitations and restrictions. This dystrophic view left little space for a more holistic and differentiated view on nature conservation. Or in other words with reference to Gorm Harste, the process did not succeed in bringing the 'needs' of the citizens and their visions for the 'future' into play.

Despite that rather negative conclusion, the national park process has been an interesting, full-scale experiment. Some important lessons can be drawn from the process and the outcome of the experiment raises the question of future possibilities. If the technical rationality of the System is not to be avoided in a democratic experiment where the democratic dimension was placed above all other dimensions, then what can be done? If the System cannot withdraw from interfering negatively in so-called democratic processes what possibilities does that leave to minimise the gap between the System and the Life World?

As described, the whole process related to the Danish national park projects was dominated by conflicts between different interest groups. Each interest group attempted, on the basis of its own rationality, to forcefully 'integrate' its political definition of 'needs' and 'future' into the rest of society. Which of the interest groups succeeded the most in so doing can be discussed, but seen from the society point of view no fundamental changes were brought about, either in relation to participation or an improved future for the biodiversity. On the local level, the result of the process was two-fold. Firstly, no common visions or solutions were created due to the power balance between the strongest interest groups and, secondly, those citizens not organised in particular groups beforehand were politically marginalised from the process. The non-thematic part of the process, however, demonstrated an interesting potential for the creation of common awareness and responsibility among those citizens actually participating, indicating that the deliberative democratic interpretation of sustainability is possible. The main lesson to be learned from that experience is the necessity to operate more explicitly with methodologies that are able to prevent marginalisation and exclusion. There is a clear need for balancing the existing power structures. The mainstream participative methodology used in recent years within environmental planning is to a large extent 'stakeholder-oriented'. In fact, issues like conflict management and governance dominate recent international literature on nature conservation. An advanced example of a governance perspective dominating the participatory process of nature conservation is to be found in the Netherlands where Maarten Hajer recounts experiences of attempts to integrate different citizens' discourses on nature into nature management discourses (Hajer 2003). Hajer's interpretation of the participatory dialogues delivers a lesson about potentials in the prevention of conflicts but not in the creation of alternative sustainable futures. The present case of Moen demonstrates the necessity to experiment with methods more capable of creating a collective orientation with a departure in a more classical understanding of people participating as *citizens* and not just *stakeholders*. The stakeholder approach will always favour the most powerful interests.

Another lesson to be learned is the importance of ownership, not just in its materialistic meaning, but also just as important in its ideological meaning. Although the thematic and the non-thematic approach described in the case did produce different levels of collective orientations, both processes were largely driven by the feeling of ownership of the land, even among those not legally in possession of the land. In that respect, a part of the problem is the fact that nature is no longer *common* as the decision-making process on natural resources has been removed from the everyday life of people. The consequence is paternalism as the personal responsibility and influence is taken away from local people and has become an issue between private landowners and governmental authorities. Nonetheless, as argued by Vandana Shiva, the revitalisation of nature as common can be seen as a precondition for a public and democratic dialogue (Shiva 2005). If nature is not a common it creates unequal preconditions for a democratic dialogue about nature, and intimidation from the usual power structures (the state, agriculture and NGOs) can continue. In its practical form, the definition of nature as a common would create a greater local responsibility and involvement and traditionalists would have to loosen their firm hold on land. The potential for participatory democracy might be enhanced, as farmers, like the planners and administrators, can relate to the Life World, even though the traditional methodology of public planning incorporates a technocratic rationality. All residents depend on social and cultural coexistence with the surrounding local community. Therefore, from a philosophical perspective, also here it should be possible to discover potentials for a new – if not completely, then approximately – way of defining sustainability.

The final question to be raised is whether public agencies and institutions are able to endorse a truly democratic sustainable development. As emphasised in the beginning of the chapter, all systemic institutions are by definition purposive. How can such systems possibly exceed their own purposive orientation? There is no clear answer to that question, but in this chapter we have been arguing for a utopian understanding of the concept that puts weight on nature and society as something that people have in common. If sustainability is to have any meaning at all, by producing changes in the way human beings coexist with nature, these changes have to be based on collective visions of the future. In that respect, it is a challenge for social scientists and others to experiment with solutions to overcome the structural barriers in the communication between citizens invited to participate in the creation of new and more sustainable visions for the future.

References

Agger, P. 2000. *Dansk naturpolitik – visioner og anbefalinger. Vismandsrapport 2000*. Copenhagen: Danish Nature Council.

Arnstein, S. 1969. A Ladder of Citizen Participation. *Journal of the American Planning Association*, 35(4), 216–24.

Clausen, L.T. 2007. *Action Research and Ethnographic Fieldwork – A Dynamic Synthesis or Two Incompatible Paradigms?* Paper to the workshop on Action Research Action Learning. University of Evora, Portugal.

Elling, B. 2008. *Rationality and the Environment*. London: Earthscan/James & James.

Habermas, J. 1976. *Borgerlig Offentlighet*. Oslo: Gyldendal Norsk Forlag.

Habermas, J. 1984. *The Theory of Communicative Action*. London: Heinemann

Habermas, J. 1996. *Between Facts and Norms: Contributions to a Discourse Theory of Law and Democracy*. Cambridge: Polity Press.

Hajer, M. 1995. *The Politics of Environmental Discourse: Ecological Modernization and the Policy Process*. Oxford: Clarendon Press.

Hajer, M. 2003. A Frame in the Fields: Policymaking and the Reinvention of Politics, in *Deliberative Policy Analyses: Understanding Governance in the Network Society*, edited by M. Hajer and H. Wagenaar. Cambridge: Cambridge University Press.

Hansen, H.P. 2008. *Demokrati & Naturforvaltning*. Ph.D. thesis. Roskilde: Roskilde University.

Harste, G. 2000. Risikosamfundets tidsbindinger, in *Dansk Naturpolitik i et bæredygtigt perspektiv*, edited by J.H. Andersen. Copenhagen: Danish Nature Council.

Luhmann, N. 1982. *The Differentiation of Society*. New York: Columbia University Press.

Ministry of the Environment. 2003. *Handlingsplan for biologisk mangfoldighed og naturbeskyttelse i Danmark 2004–2009*. Copenhagen: Ministry of the Environment.

Ministry of the Environment. 2003. *Letter of Approval Sent to Selected Pilot Projects*. Copenhagen: Minister of the Environment.

Mol, A.P.J. 1996. Ecological Modernisation and Institutional Reflexivity: Environmental Reform in the Late Modern Age. *Environmental Politics*, 5(2), 302–23.

Nielsen, K.A. and Nielsen, B.S. 2007. *Demokrati og Naturbeskyttelse. Borgerne i Nationalparkprocessen*. Copenhagen: Frydenlund.

Nyord-foreningen, 2004. *Nyord som del af en nationalpark – er det nu en god idé?* Nyord: Nyord-forening.

OECD. 1999/2000. *Environmental Performance Reviews: Denmark.* Paris: OECD.

OECD. 2000. *Denmark – Environmental Performance Review (1st Cycle): Conclusions & Recommendations, 32 Countries (1993–2000).* Paris: OECD.

Sachs, W. 2000. *Development: The Rise and Decline of an Ideal.* Wupperthal papers, 108. Wupperthal: Wupperthal Institute for Climate, Environment and Energy.

Shiva, V. 2005. *Earth Democracy: Justice, Sustainability and Peace.* London: Zed Books Ltd.

Tind, E.T. and Christensen, H.S. 2001. *Nationalparker i Danmark: En diskussion på baggrund af udenlandske eksempler.* [*National Parks in Denmark: A Discussion Based on International Case Studies.*] Copenhagen: Danish Nature Council.

UNEP. 2000. *The Convention on Biological Diversity.* Montreal: Secretariat of the Convention on Biological Diversity.

Wilhjelm Committee. 2001. *En rig natur i et rigt samfund.* Copenhagen: Ministry of the Environment.

World Commission on Environment and Development. 1987. *Our Common Future.* Oxford: Oxford University Press.

Chapter 12

Health, Food and Sustainability

Kirsten Bransholm Pedersen and Birgit Land

Introduction

There is an apparent paradox: there has been an increase in diseases related to lifestyle and environmental problems, while at the same time there has been a growth in the number of initiatives undertaken at both national and international levels to develop strategies for the promotion of sustainability and health respectively. How can this be? Our answer to this question is that the two types of strategy have not been integrated. This is despite the fact that the problems often have the same cause, namely humans' productive transformation of nature resulting from an increasingly intensified process of industrialisation.

This separation is not restricted to government bodies. It also applies to education and academic research and their institutional contexts. Different disciplines have been placed under different ministries and in different departments at universities resulting in a lack of cross-disciplinarity, and the ignorance of potential overlaps. In some cases, this has even resulted in the creation of new environmental or health problems. Examples include the cultivation of rainforests. Here the intent has often been to solve health problems and provide supplies of food, and the negative side effect has been the exhaustion and erosion of great areas of land; individual strategies in the Western world to stay fit and healthy, for example by eating fish with omega-3 essential fatty acids, have created the basis for the extensive fish and shellfish breeding industry in, for instance, Thailand resulting in extensive destruction of lakes and swamp areas; governmental and supranational initiatives to preserve biodiversity by converting forest and areas of swamp into national parks have deprived local populations of their basis for existence, not surprisingly producing a critical attitude towards initiatives aimed at sustainability.

Thus the aim here is to integrate these two issues. We argue that understanding sustainability strategies via health entails relating sustainability to social- and local-based conceptions of 'the good life' based on experiences and life expectations. Understanding health via sustainability will intensify the attention given to society and democracy. Consequently, the discussion of health promotion will be elevated from the individual level, which is the current emphasis, to the public and collective level where, in our opinion, solutions to health problems should be sought. Consequently, in theoretical terms it becomes essential to discuss and reassess the concepts of health and sustainability, and the positive synergy which will be achieved by thinking these concepts together.

Our argument will be illustrated by a *study of food production and consumption*. The production and supply of foodstuffs is central to our understanding of the growth in health problems. This is because supply of food forms the frames of possible consumption both in terms of quantity and quality, whereas the food we can buy must be seen as the result of a long journey throughout a complex food chain on which the consumers have limited influence. At the same time, food production, including agricultural production, can have negative environmental consequences which may also cause health-related problems.

Furthermore, it is necessary to understand the change in meaning of food consumption that has taken place in late modern society. The everyday life approach employed here places both people and their consumption in their social and cultural contexts, hence the analysis will include the examination of the conditions of consumption, which have consequences for both health and the environment.

An important aspect of this chapter is to discuss which choices of possibilities and which strategic interests different groups of actors have within the field of 'health and sustainability'. Here we will advocate that a shift from the individual to the societal perspective requiring a more active government regulation in favour of a shift in the politics of food and health is desirable.

Health and Sustainability: Conceptualisation

The discussions on *health* and the new diseases, which have surfaced in our global, late modern societies have captured a central position on the political agenda. At the same time, the discussion on how to understand health as a scientific concept has intensified. The concept of health and health promotion is discussed within the social sciences, natural sciences and humanities, for example within sub-disciplines such as medicine, ecology, sociology, biology, psychology, political studies and economics. Hence health is a multidisciplinary concept and, as we want to emphasise in this chapter, it is becoming a vital component in our understanding of how to ensure sustainable development in the future.

The Brundtland Report, within the organisation of the UN, introduced the concept of *sustainability* in 1987 and emphasised the relation between development, nature and the environment. However the Report underlines that for development to be sustainable it must include aspects of social and economic sustainability. As part of social sustainability *health* is mentioned in the Report. For development to be sustainable, it must meet essential human needs. Human needs are described as: housing, water supply, sanitation and *health care* (World Commission on Environment and Development 1987: 55).

As early as 1946, the World Health Organization (WHO) defined health in positive terms as 'A state of complete physical, mental, and social well-being and not merely the absence of disease or infirmity' (WHO 1946).

In practice, however, the definition has been implemented in a more narrow sense and the question of a good health system has influenced the policies of WHO. For example, in WHO's 2000 World Health Report, health attainment in 191 member states is measured by concepts of 'a good health system' and access to services. Distribution of health in the population is included (WHO 2000).

Along the same lines, WHO in the 1980s defined the concept 'environmental health' as comprising those aspects of human health and disease determined by factors in the environment. In this document environmental health problems are conceived as related to the physical interaction of environmental and health factors. Later, in the 1990s, WHO's approach to the concept of environmental health was widened to encompass social and psychosocial conditions which have an effect on health, arguing that even if it can be difficult to prove a direct correlation between some of these social factors and health, it should not make them any less important in the consideration of environmental health priorities (WHO 1998: 8). Consequently, WHO defines environmental health as comprising:

> [T]those aspects of human health, including quality of life, that are determined by physical, chemical, biological, social and psychosocial factors in the environment. It also refers to the theory and practice of assessing, correcting and preventing those factors in the environment that can potentially affect adversely the health of present and future generations. (WHO 1998: 9)

This approach implies that improvements or solutions to the problems have to be found in the elimination of physical, chemical or biological risks in the environment and in the improvement of the social and psychological environment. Put into practice, WHO proposed that environmental *health services* must be established as services implementing environmental health policies through monitoring and control activities (WHO 1998).

On the basis of data from WHO, Lang and Heasman have provided a good understanding of ways in which WHO sees environmental factors affecting health (see Table 12.1).

Compared to the comprehensive definition given above, factors other than physical and biological ones almost disappear. Besides physical factors, only unhealthy housing is stated to cause ill health. Psychosocial factors seem absent.

Within recent years the tone of the debate on sustainability, and not least on the important part health plays in that, has been sharpened. While in the 1980s the concept of sustainability still had to be argued for as a relevant scientific category when discussing development, today, in the Medical Journal of Australia Nick Towle concludes that '*The ultimate consequence of failing to live sustainable is that we push our own species to extinction*' (Towle 2004). With species he refers to both humans and non-humans. Devastating health effects and environmental degradation have pushed us closer to The Edge, he argues.

Table 12.1   Environmental factors affecting health

	Polluted air	Poor sanitation and waste disposal	Polluted water or poor water management	Polluted food	Unhealthy housing	Global environmental change
Acute respiratory affections	X				X	
Diarrhoeal diseases		X	X	X		X
Other infections	X	X	X	X		
Malaria and other vector-borne diseases		X	X		X	X
Injuries and poisonings	X		X	X	X	
Mental health conditions					X	
Cardiovascular diseases	X					X
Cancer	X			X		X
Chronic respiratory diseases	X					

Source: Based on Lang and Heasman (2004); UNEP (2000).

Health at the Forefront

Tony McMichael, Director for the National Centre for Epidemiology and Population Health, The Australian National University (McMichael 2002) brings a new angle into the discussion on the relations between health and sustainability. He argues that we can no longer maintain the health and well-being of human population if we understand health merely as an incidental benefit of 'sustainability'. Rather it should be regarded as a central target or criterion. He wants to see health come to the forefront in the discussion of sustainability.

Thomas, Douglas and Cohen are also critical of WHO's understanding of the relationship between health and sustainability forwarded by WHO and the Brundtland Report, which only include health in the terms of health services or health care. They suggest a conceptual distinction between 'sustainability in health' and 'sustainability and health' (Thomas, Douglas and Cohen 2002). The first issue concerns health as an industry, mainly focused on health services, with its own resource requirements, economically beneficial outcomes and potential for environmental pollution. The main concern *in* health is the ever-increasing resource demand, therefore they argue that this apparently never-to-be-satisfied consumer pressure for health services is coupled to an ever-increasing resource demand, and hence potentially, a rapid and non-sustainable economic activity.

The second concept 'sustainability *and* health' offers a broader conceptualisation. Here health is seen as a central focus for all other sectors of sustainable or non-sustainable development. Health is on the one hand the outcome of material (sustainable or non-sustainable) production and on the other, a condition for sustainable development. Seen from the last perspective, health must be regarded as a positive contributor to sustainable development seen from both an economic and a broader social and societal angle.

Below, with the aim of contributing to the ongoing discussion on how to understand the interaction between health and sustainability, and to developing politics focusing on the complex interactions between them, we find it relevant to take a closer look at more recent scientific contributions concerning how to understand and bring into play the two concepts in new ways.

What we find interesting and relevant in the works we will present is the understanding of the concepts as not primarily instrumentalist tools, but rather as a normative horizon for orientation in late modernity. Moreover, the concepts provide a theoretical and practical contribution to how to understand and discuss both the responsibility and the possibilities of the individuals and to discuss responsibilities at the societal/political level.

Sustainability: A Normative Horizon for Orientation

The last 20 years, since the Brundtland Report was published, has seen many discussions, interpretations, understandings and operationalisations of the concept of sustainability. At the institutional level the prevailing way of operationalising the concept has been to set levels and directions of development through a system of indicators in an attempt to make monitoring and control possible.[1] A radical criticism of Brundtland's concept of sustainability has been put forward by Sachs, Loske and Linz (2000) and by Olsén, Nielsen and Nielsen (2003). The essence of the critique is that the concept as it has found use does not free itself of a growth-based concept of development. On the contrary they argue the growth-based development of society has absorbed the concept of sustainability, which is used as a technocratic and instrumental tool for pure economic growth. For this reason, Sachs, Loske and Linz (2000) argue that the concept is useless. Other researchers, e.g. Olsén, Nielsen and Nielsen (2003), although also critical to the technocratic reading of the concept, have been proponents of keeping the concept, and make use of it as a normative horizon for democratisation and equality. Olsén, Nielsen

---

1   The Wuppertal Institutes in Germany has been a central actor in this kind of research.

and Nielsen (2003) still see a potential in using the concept of sustainability for reaching qualitative goals for societal wealth and welfare. Coming to a point where the concept of sustainability can give answers to how to reach those goals, the way we produce knowledge concerning sustainability becomes essential. They draw attention to the fact that the instrumental interpretation of sustainability leads to distinct and limited production of knowledge. Abstract knowledge of experts regarding facts about, for example, increasing CO_2 emissions, climate changes, and chemicals in production and goods is increasingly drawn upon when the global goals for sustainable development should be reached through actions and strategies at the local level, in national political strategies, in companies, institutions and in everyday life. New and better knowledge about these conditions are of great relevance, but reliance on expert knowledge leads to a disconnection of knowledge based on experience, which is something that has existed in all cultures – a knowledge about the use of nature in relation to the production of daily necessities. When knowledge is disconnected from the praxis of everyday life, it has a great impact on the individual process of formation and increases the insecurity of orientation. Therefore they offer an alternative understanding and implementation of the concept of sustainability:

> ... we will plead for a critical project of enlightenment where humans' reason and self-understanding are reflected in relation to nature – in a way of living, which includes democratic participation in shaping production, the living environment, family life, consumption and social and cultural societal life. Only by opening up the possibility that humans themselves bring about social responsibility and sustainable development – e.g. during democratic production – will a sufficiently rooted responsibility, which considers the reproduction of the resources of nature, aesthetics and biodiversity as a part of the respect for intrinsic value of nature. Such a responsibility is included in the dawning horizon for a sustainable development. (Olsén, Nielsen and Nielsen 2003: 265–6)

Understanding sustainability in this way will emphasise the division between the environment and the social effects of economic growth enabling a positive discussion of wealth, justice, equality and welfare (Olsén, Nielsen and Nielsen 2003: 264).

It is this broad and critical understanding of sustainable development that we want to bring into use and to develop further in relationship to the broad notion of health.

Like the concept of sustainability, the concept of health during the latest years has been subject to theoretical discussion. Recent contributions within research in health studies have focused on and point out how a broader understanding of health – and health policy – can be an important precondition for how the individual can see the meaning and is able to present sustainable and healthy actions in their everyday life.

How to Understand Health in a Wider and Holistic Perspective

As previously argued for, we suggest a wider understanding of health than that of health care and absence of disease. Realising the complexity of our contemporary society, where everyday life and working life have changed radically, and new diseases resulting from one's individual lifestyles have also changed, health related to sustainability must be seen in a wider and more holistic perspective.

If we look at the *practice* of WHO instead of its definition, as we mentioned earlier, it would be obvious to understand health as *absence of disease* at a personal level. This understanding also corresponds with the common understanding of the concept that many people have and, more importantly, it is in line with the bio-medical concept of health, which has been implemented and practised in the Western world since the Age of Enlightenment. However, partly under the influence of the growing number of lifestyle diseases, to which it has been difficult to produce satisfactory medical solutions, the critique of this rather narrow definition, with a view of human nature embedded in biological sciences, has been intensified. Today, the thinking we find in *sustainability and health* is establishing a position for itself in both lay and scientific discussions on health. Practitioners from different occupations and researchers from the humanities and social sciences representing different scientific approaches have offered alternative interpretations of the concept of health. Different perspectives are introduced but we find a general agreement that for health to be sustainable it must be comprehended as *something more than* absence of disease.

Health: A Question of Mastering One's Life

Elsass and Lauritsen suggest in their book 'Humanistic Health Research' (Elsass and Lauritsen 2006) how to systematise the different conceptualisations of health. As their starting point, Elsass and Lauritsen outline different conceptions of the causes *of disease*. Here they point out two highly different views, the biomedical scientific paradigm, which is today 'the established' paradigm opposed to this 'the alternative'. The alternative contains a wide range of contributions from sociological to psychological and bio-psychological/holistic views (Elsass and Lauritsen 2006: 38–9).

The biomedical approach, where causality, which means correlation between cause and effect, is a key concept, is embedded in the positivist scientific position. The causes of diseases are looked for in recordable disorders (disturbances) in the functions of the human organism. Health here becomes absence of disease, and health is obtained by removing or reconstituting disturbances which have caused imbalances in the organism.

The critics of this approach have referred to it as 'the machine defect method'. This metaphor expresses the criticism that the human organism is understood as a machine which by functional disorder again becomes functional by mechanically replacing or repairing those parts.

Critics from 'the alternative' approach to health understand health (and disease) in a broader context. Elsass and Lauritsen mention that the point of departure of this approach e.g. could be 'life quality in balance with nature', where the focus is on 'a healthy life' and disease arises when a person does not live, or is not given the possibility to live, 'the good life'. The crucial difference is that the focus for the identification of causes has been moved from the body to broader psychological or societal conditions for life. Subsequently, 'the alternative' approach ascribes greater importance to the idea of prevention and strategies for sustainable development at the societal level than does the bio-medical approach.

In one of the most often quoted scientific articles the philosopher Steen Wackerhausen, who is a representative of one of the alternative approaches (Elsass and Lauritsen 2006), defines health as:

> the contextual capacity of acting, which more specifically is an expression of the relation between the individual's goals (values), its conditions of life and its bodily tied (psychical and physical) capabilities of acting. (Wackerhausen 1994: 43)

According to this definition, health is a means to achieving valued goals circumscribed by an individual's circumstances.

Antonovsky (1987), like Wackerhausen, has focused on the circumstances which enable positive health rather than on the conditions that make people sick. From a socio-psychological point of view he stresses that strategies for health promotion must involve a dynamic way of thinking, where experience, recognition, processes of learning and inner feeling of coherence, are of similar importance as conditions of life and can be subordinated to different stress factors (Antonovsky 1987, Jensen and Johnsen 2000).

Philosopher and health researcher Juul Jensen takes up a critical position to the vital importance of the above and the bio-medical science paradigm to health attached to the capacity of the individual. He argues that 'the collective', understood in a broad meaning as family, football club, work place, etc., must be involved and be the basis of the discussion on health promotion. Social collectives must 'learn' to take care of each other and ensure the individual is capable of participating in social life (Jensen 2006).

Elsass and Lauritsen (2006) pay attention to one more perspective – namely, the societal, comprehended as societal discourses. That means the way we possibly *can understand* and discuss health. Following Elsass and Lauritsen, we need an approach and a methodology from which we critically can deconstruct the paradigms and interests represented when we are talking about health promotion in our contemporary society. This perspective and critical view is shared by the sociologist Dorthe Gannik (2005). However, for her, it is important to emphasise the meaning of context including care in close relations and working conditions. Also, she emphasises the importance of analysing the constructions of disease and health which are produced and reproduced in those situated contexts. To Gannik disease is something 'born out of' the social. On the basis of this she concludes that knowledge relevant to transforming the context and to promoting individual and collective action must mainly be related to the social context and to the room for manoeuvre given to the individual to act sustainably, rather than bio-medical knowledge about the individual as a carrier of disease (Gannik 2005: 337).[2]

The above contributions from the field of health research in sociology and psychology bring into play several new perspectives on health, which are relevant if health promotion should support sustainable development and 'vice versa'. One substantial point which can be drawn is that health cannot be provided (solely) at an individual level. Societal conditions for producing health constitute the frames for individual action. These include working and environmental conditions, knowledge, the economy, and social and cultural understandings of health. We will also argue that health strategies to a higher degree must be disconnected from the market and the dynamics behind economic strategies of growth. For example, the individual can buy healthy food, but cannot buy a healthy environment, the individual can buy industrial accident insurance, but cannot buy a healthy working environment, and the individual can buy an energy friendly car, but cannot buy a sustainable transport system. In total, politics and societal sustainability constitute the frames for individual health strategies.

Another relevant perspective brought into the discussion is the question of *interests* involved in health promotion. The state needs the population to remain healthy in order to avoid the explosion of public health expenditure and to keep a healthy work force. Individuals on the other hand, might to a higher degree associate health to their normative orientation towards 'a good life'. Therefore to obtain sustainability and health, the point of departure should not be an instrumentalist approach to keeping the population healthy for economic reasons. If healthiness is to become 'the people's project' and hence sustain both individual healthiness and a wider societal economic, social and environmental sustainability, strategies must emerge from and be coupled to the normative and political discussion on how to keep and improve quality of nature and human life.

The Duality of Health and Sustainability

If we return to McMichael (2002) and Thomas, Douglas and Cohen (2002), and the discussion on how to conceptualise the relation between sustainability and health, we could add health as a fourth dimension 'in its own right' to the concept of

---

2   The broader concept of health is highly contested within these years. One critical argument is that the many dimensions influencing health seem equally important, which is why it is almost impossible to use the concept as a guideline for practical interventions. In this chapter we will not contribute to this discussion, however we want to stress that asking for simplistic, rational solutions might blow the absolutely necessary discussion on how to understand health and solve health problems today, back to the time where the bio-medical approach made up the entire frame for understanding.

sustainability – that development to be sustainable now comprises environmental, economic, social and health sustainability. However, to underline the above points of giving the concept of health and the concept of sustainability new meaning, we want to promote a slightly different conceptualisation of the relationship between health and sustainability. Inspired by Anthony Giddens (1984), we want to suggest that *the duality of health and sustainability* is chosen as a starting point for understanding the two concepts, their mutual impact and the 'demands' they put on each other to foster a healthy and sustainable future. Anthony Giddens, with his theory of structuration, stresses that the constitution of agents and structures are not two independently given sets of phenomena, a dualism, but represent a duality where agents and structure are seen as mutually enabling and equated with constraints (Giddens 1984). Likewise, we find it fruitful to conceive both health and sustainability as mutually enabling and equated with constraints. They mutually produce, reproduce or destroy each other.

Figure 12.1 illustrates that thinking along the lines of a duality between sustainability and health has consequences for the way we understand environmental, social and economic development (see Figure 12.1).

By understanding health and sustainability as a duality, health both creates conditions and is conditioned by sustainability, understood as economic, social and ecological sustainable development, while on the other hand sustainability creates and is conditioned by human health. In other words, health promotion must be integrated more actively into the concept of sustainability and sustainability in the concept of health.

This conceptualisation implies that the demands The Brundtland Report makes on development to be sustainable must be extended and stated more precisely. Health and sustainability must be conceived as a duality in which each creates and conditions the other. *Sustainability must be conceived in a health perspective and health must be conceived in a sustainability perspective.* Consequently, health promotion must take sustainability into account. Failure to do so will be discussed later in this chapter.

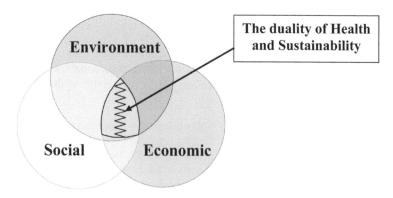

Figure 12.1   The duality of health and sustainability

The Duality of Sustainability and Health: The Case of Food Production

As already mentioned in the Introduction, we will take the case of food production and consumption to illustrate the duality of sustainability and health. For analytical simplicity we divide sustainability into environmental, economic and social aspects, and ask how the concept of sustainable food production and consumption must be conceived to meet the demands of producing health, and how the concept of health must be conceived to meet the demand of sustainable food production.

The Environment: Sustainability and Health

Sustainable development according to the Brundtland Commission has to do with protecting nature and the resources of nature. Integrating a health perspective, however, means that sustainability also includes the body. Therefore from the perspective of duality we must put a demand on food production to guarantee nutritionally safe goods, e.g. free of chemicals. Furthermore, industrial pollution which has a negative impact on health must be prevented to ensure a healthy reproduction of both nature and human beings.

Conversely, if the point of departure is improving the *health* conditions by mean of food production, this can result in severe environmental problems if only focusing on the one side (health) of the dual relation.

Today we are witnessing intensified food production to support the increasing population of the Third World with food they can afford. The Green Revolution can be taken as an example of that. This programme was initiated by the United Nations (UN)/Food and Agriculture Organization (FAO) to increase food production in developing countries and hence ensure the supply of food for the poor people in these countries. It is disputed if the goal of this programme has been attained, but many politicians and scientists from developing countries, among them Vandana Shiva, have, from a sustainability point of view, severely criticised this policy. In India, modern agriculture employing pesticides and other inputs was introduced, and traditional sorts of grain were substituted by high yielding but barren hybrid varieties. The consequences were loss of biological diversity and resources, and ecosystem destruction associated with industrial and commercial development (Shiva 1989: 145, Shiva 2006). If this development is seen in the duality between health and sustainability, and if the farmers involved – with experience-based knowledge and provided resources –had been included in the process of development, then the process probably could have been detached from the market's demand of growth and yet ensured enough food for the poor in the developing world at the same time as the cultivation of crops took place in such a way so that the land was not exhausted and yet produced healthy food and local jobs.

The Economy: Sustainability and Health

Discussing market and economy, the Brundtland Commission sees global growth in food production as a precondition of improving the standard of living and hence health conditions in the South. It is pointed out that two conditions need to be secured, namely that the sustainability of ecosystems on which the global economy depends must be maintained, and that the basis for exchange between economic partners needs to be equal which is not the case today (World Commission on Environment and Development 1987: 67). However, when we take health, equal access to food and democracy into consideration the distribution of wealth in society must also be reconsidered. Sufficient supplies of food on the market do not guarantee health. Food must be offered at prices which ordinary people can afford, and the globalisation of the agro industrial business has to be evaluated in terms of the impact it might have on poor people's possibilities for local self-sufficient food production, and thus their conditions for living a healthy life. If we use the fish industry as an example, FAO estimates that while both the number of fishermen and the catch have grown in the export-oriented fishery in Western Asia, 23 million remain income poor, working either as fishermen or in related jobs (Lang and Heasman 2004: 244). This case illustrates that development policies aimed to supply both North and South with healthy food and wealth, have failed in the South, where both the environment and the poor people have paid the price for supply of fish to the wealthier countries.

The traditional concept of sustainability deals with growth as a means to ensure health, also in the South, but disregards the fact that production and distribution of the growth take place under market conditions. In many cases this means that neither the objective of health nor sustainability will be met. If nature is to be taken care of we must prioritise a socially sensible use of the resources of nature, where regulation must aim its attention at the well-being of nature and health, and a decoupling from the market mechanism.

Consequences of the Development of Food Production

Lang and Heasman (2006) argue that the global environmental and health problem we are facing today, which is related to food production and distribution, has its roots in the dominate paradigm which has been behind Western food production strategies since the Second World War.

After the Second World War, great attention was drawn to the lack of many different goods, starvation and undernourishment. The response of both the farming and food industries was the fundamental industrialisation of food production. Lang and Heasman (2006) call this the *Productionist Paradigm*. The Productionist Paradigm was based on the widely accepted societal rationale for ensuring that the health and nutritional needs of the population were met. Increasing production and making it more effective required both the industrialisation of production and monoculture rather than diversity. This in turn entailed the use of fertilisers, pesticides and energy intensive machinery.

The focus on the quantity of production rather than quality and the pressure of global world trade on local markets and living conditions have simultaneously caused an increase in the unequal distribution of food and health between developing countries and the Western world. It is now possible for the wealthy part of the population in the West to choose between an extensive supply of fresh and exotic products all the year round, just as the consumption of meat has grown within this group. The situation is different for many developing countries. Lang and Heasman (2006) write that in spite of the historical emphasis on improving national self-sufficiency and food production in the South, globalisation and changes in the Western diet have ironically led to a deterioration of national self-sufficiency in many developing countries. The result is growing malnutrition globally: undernourishment in many developing countries and over consumption and obesity in the Western world (Popkin 2006). In the light of the duality between health and sustainability, the consequences of this way of producing food can be seen both as a deterioration of the resources of nature and human health.

Thus, the concept of health within the Productionist Paradigm is a relatively narrow concept that falls within the bio-medical understanding of health and limits itself to a question of the production of enough food and has therefore fitted in well with the interests of the food industry and the *growth* in production and the economy. But, as we have described, this quantitative expansion and intensification of production have had serious environmental consequences. The increased attention to these issues has resulted in strategies both at national and international level, whose target has been to reduce the negative environmental consequences. One example is the preservation of great areas of forest in the form of nature parks in developing countries where the purpose has been, for example, to maintain biodiversity. However, the problem has been that this strategy did not include the broad notion of health. The projects have in many cases led to the loss of the livelihoods of the local population, and they survive, inter alia, by poaching in nature parks or by seeking jobs in the cities where slum areas are an increasing problem. Therefore we agree with Lang and Heasman (2006) that it does not make sense anymore to argue for the Productionist Paradigm in the discussion about a sustainable future and health promoting development (Lang and Heasman 2006: 35–7).

There are several competing alternatives to the Productionist Paradigm which claim to improve both the environment and health. Lang and Heasman (2006) assume that in the future, the food economy will rely more on biological than on chemical science, and they consider the medical-biological/Life Science Integrated Paradigm and the Ecologically Integrated Paradigm as the two most realistic emerging alternatives.

The Emerging Paradigms in Food Production

One strategy to prevent the negative consequences of the Productionist Paradigm is embedded in the *Life Science Integrated Paradigm* (LIP). The application of biotechnology to food production on an industrial scale is the core of this paradigm, and techniques found in biotechnology are already delivering many new food production methods. *Genetic modification* (GM) has become the central defining characteristic of LIP. GM is today used in the world food system at a rate that some see as irreversible. From the mid-1990s to 2006 the global cultivation of GM crop production had rapidly risen from zero to 102 million hectares (ISAAA 2006).

Nutrigenomics is another upcoming research field within this paradigm. Here, researchers are investigating how genes operate in plants, animals and human beings, to understand how nutrition and particular dietary intakes interact with individual people's genes. Many experts see the LIP paradigm as an almost perfect solution to the many problems of the Productionist Paradigm, from lessening environmental problems to improving human health through increasing food production and producing new kinds of food with higher health benefits.

However, critical voices argue that nutrigenomics is an extremely individualised policy approach, and only the 'worried well' and rich consumers might gain benefits from this research (Lang and Heasman 2006: 22–4).

Moreover, both democratic and environmental concerns have arisen. One concern is that 'big business logic' is undermining the public ability to make informed choices and, hence, in terms of the health conceptualisation of Wackerhausen, deprive people possibilities for mastering their lives.

Lang and Heasman moreover argue that the concept of health within this paradigm is bound up with the bio-medical approach, and food is perceived as almost a drug, and part of a planned, controllable and systemic manipulation of the determinants of health and ill health (Lang and Heasman 2006: 22).

The organic movements have criticised the LIP. They point to *the Ecologically Integrated paradigm* (EIP) as an alternative. The EIP is, as is the Life Science Paradigm, grounded in the science of biology, but interprets biological and social systems in ways that offer different choices for our food and health future. Lang and Heasman see EIP as a positive and the only sustainable alternative to the Productionist Paradigm. EIP takes a more holistic and less engineering approach to nature and a more holistic view of health and society than does the 'medicalised' approach of the Life Science Paradigm.

Summing up, we will make an appeal to implicate the duality of sustainability and health in the development of food production in the future averting the most critical consequences for the environment and health.

In the next section we will examine the social dimension of sustainability and health by casting light on the possibilities and the barriers of the consumers to play a significant role concerning health and sustainability.

Sociality: Sustainability and Health

If we take a look at the social perspective, then it is not enough to produce sufficient food, nor is it enough to distribute it equally. Malnutrition today is a growing problem both in the South and in the North, and regardless within which of the above-mentioned paradigms food could be produced this problem will arise. One reason for this is that humans are not necessarily rational in relation to the environment and nutrition in their consumption of food. Individual choices are entangled in the structure and the practices of everyday life, and whether or not the consumption of food will go in the direction of sustainability and health will depend on habits and reflections on how to make up the most satisfactory everyday life with many conflicting goals at various stages, and within the social and societal frames given.

Therefore, the demand for health and sustainability also implies a focus on how the consumption of food is organised in everyday life: knowledge, economic power, and how we construct pictures of healthy and sustainable practice are at this stage.

In everyday life, the consumer makes a series of choices about what to eat, how to make the food, and in which social context it should be ingested. Giddens (1991) characterises this as a part of a life policy. A life policy is a policy where the individual makes a series of reflective choices based on experiences and information. These choices all have a more or less health promoting or environmental meaning, but they can also in themselves imply health promoting strategies or strategies for improving the environment. Numerous examples of this can be mentioned, namely to incorporate physical activities in everyday life, biking instead of taking the car, to buy organic food, pre-sorting of waste, to live in healthy surroundings, to eat healthily, and not to smoke. Some of theses actions will in themselves have a health promoting function, e.g. going for a run, others will only have an environmental effect, such as the pre-sorting of waste. Many actions in everyday life will have both an environmental and a health promoting effect, e.g. the consumption of organic food, eating more vegetables and less meat, and taking the bike instead of the car. Those reflective choices in life policy are primarily built on experience-based knowledge that is tied to strategies for 'the good life', but is also constantly confronted with the knowledge of experts on the dangers and risks of, e.g. the content of heavy metals in fish, or the destruction of mangrove swamps caused by shrimp farming. This illustrates that different and sometimes competing forms of knowledge are at stake. Often *health*-related practices will be based on a physically-based experience and lay knowledge while considerations such as whether the production of food has been *sustainable* will often be based on expert knowledge.

If the dimension of sustainability could be more closely attached to considerations of health and to a normative horizon concerning sustainability, we suggest that this dimension could be more successfully integrated in choices, which are taken in everyday life practices.

Moreover this shall be seen in relation to the fact that the dimension of health and sustainability in the consumption of food in late modernity plays a limited role because of strong competition with other identity creating factors, new rhythm of time and other social factors in everyday life (Halkier 1999, Land, Pedersen and Ilsøe 2007).

New Rhythm of Time: New Meals

When discussing healthy and sustainable food practices, we find it important to take into consideration the changed societal settings in the Western world, which has led to structural changes in everyday life in the form of changed family patterns, working life and working hours, housing conditions, patterns of mobility, women's work outside the house and so on. This has also changed the habits of food and patterns of diets and thereby the consumption of food. The structural changes in everyday life can consequently, as with Gannik's (2005) understanding of the structural changes of working life, provide barriers for the space of action, in this sense, e.g. the creation and eating of meals together, these are replaced by fast food. This can be a factor which hinders the creation of an environmental and health promoting culture of food.

However, we do not see this trend as unequivocally negative. We also see some potentials. Decisions that formerly were made at the level of the individual or family have now often 'moved out of the house' to societal institutions such as schools, hospitals, sports clubs, nursing homes and work places, where experts such as catering officers, chefs and dieticians are given the possibility to take over responsibility and supply healthy food. These choices can be positive or negative regarding the environment and health, and specify new demands for a public supply and regulation of the area of food.

Summing up, we find that individuals *have* a room for manoeuvre when it comes to eating healthily and sustainably, moreover that this room has been both extended and limited during the process of the modernisation and industrialisation of food production. In Western countries most people have access to food, and know what is healthy to eat. Also, but only to some degree, they know about the environmental consequences of their food choices, however what is put on the shelves in the shops is to a great extent in the hands of producers, suppliers and politicians. Furthermore, the disembedding of meals from the family to public institutions, as mentioned above, restricts the choice of the individual consumer. To understand consumer choices and to discuss possible changes, the broad conception of health understood as possibilities given for mastering one's life must be involved, and public discourses and strategies on health and sustainability must be reconsidered.

In the final section we will discuss which demands the fact that the predominant part of the food production takes place within the frames of the market economy puts on regulation at a societal level. The possibilities to emancipate the production from the market fixation – which can only happen to some extent and in small steps

– will be seen as a key issue if the production should take place in a sustainable manner and reach health-related goals.

Needs for Shift in Food and Health Policy

Policy making is essentially a social process; it is a battle of interests on many levels. Understanding how food policy is produced, actors in all links of the food supply chain must be taken into consideration; from farmers, producers and retailers to consumers. In the previous parts of this chapter we have discussed potential actions at the level of food production and at an everyday level. However, as Lang and Heasman make us aware, health is not seen as a prime responsibility of any one group in the chain, and therefore it can fall down in the gap between sectors (Lang and Heasman 2006: 16). Health has been marginalised in the food economy. The attention of the Brundtland Report to the environment has caused some positive impact on choice of technology, emissions, etc., but as we have problematised, only some aspects of environmental sustainability have been focused on. Moreover, that the dual relationship between sustainability and health has not been taken into account and that the bio-medical approach to health has been the dominating discourse means that the overall goals of sustainability and health have not been met. Therefore in this part we will focus on the role food governance or policy carried out on national and supranational levels have to play and include the discussion about public democratic control and public responsibility.

On the supranational level, WHO demands a food policy that goes beyond physical and social health but also implies sustainable food delivery. Most countries support the overall guidelines towards a sustainable and health-related development of food production and distribution. Nevertheless, the policies of the individual countries are rarely directed at the achievement of sustainable food supply and limitation of the supply of unhealthy products. In this context economic interests are at stake and political actors do not seem to contest strong economic interests (Lang and Heasman 2006, Ilsøe 2006: 102). Heasman and Lang provide the example of the large producers' soft drinks that influence political decisions in the direction of maintaining the huge consumption of sugar and go against levies on sugar. However, some examples of policies which are implemented, safeguarding the health of the population, do exist despite the fact that they go against the interests of capitalisation. One example, which, however, is situated outside the actual area of food, is the prohibition of smoking in restaurants and public institutions together with requirements of labelling packages of cigarettes as being dangerous to health. This example could lead the way towards a more restrictive policy in the area of food in the future where the market is regulated with, e.g. levies on sugar and fat and by a prohibition of the products most dangerous to health, e.g. colouring agents in candy and other foodstuffs, or by prohibition against advertising these products.

If we turn to public governance aiming to regulate or support consumer strategies for living healthily and sustainably in everyday life, we have to call attention to the fact that campaigns and nutritional advice aimed at advising individual people on how they can take better care of themselves have failed as strategies to reduce or prevent ill health. The *context* for living healthily and sustainably must come into focus. For too many people the *choice* to live a healthy life is not a real one. Politics must also be addressed at the institutional and societal level. The processes of globalisation and industrialisation have in some ways made it more complex and difficult to meet the problems concerning living a healthy and sustainable life. Worth mentioning is, however, that this process has opened up new societal organised ways of supporting healthy and sustainable ways of living and eating. Meals consumed outside the home, in public institutions, in workplaces, etc., have become a possible new field for supporting a healthy and sustainable food culture. Yet this change in public governance can have both negative and positive consequences. On the one side we have already seen examples of food production in catering centres, e.g. in hospitals, where principles of industrialisation and economic rationales have been adopted which do not leave much room for goals of quality, sustainability and considerations of health. On the other side, the large-scale operations already provide an opportunity for public institutions to act as an influential consumer on the market and thereby stipulate demands about the supply of healthy and environmentally friendly foodstuffs. Finally, public meals provide a greater opportunity for a political steering in the field, which opens up for initiatives in the direction of the organisation of meals, where sustainability and health might be included. Examples of this are several municipalities' development of organic lunch programmes in schools or many workplaces' development of organic company lunch programmes. But when no security for the incorporation of sustainability and health in the institutional organisation of the meals exists, it is necessary to stipulate the demands of such a food policy in public institutions at the societal level, for instance through an elevation of the nutrition and environmentally-related quality of meals in schools and institutions, and an involvement of the users in the production when it is possible. In particular, the state in some fields has the possibility to detach itself from the markets rationales of growth and instead focus on qualitative demands for production through a closer link between producers and users.

Concluding Remarks

This chapter took its point of departure in the growing problems related to both human health and the environment. We have drawn attention to the fact that strategies for finding a solution to the problems of health and sustainability have not been sufficiently interlinked, even though these problems are often a consequence of the same development strategy. Here we referred to the industrialisation of agriculture in the developing countries, which partly solve the famine but made the environmental problems worse and to the fish industry, where poor fishermen in the South have paid the price for supply of fish to the wealthier countries.

We have thus shown that a lack of duality might have made matters worse. Taking our starting point in a discussion of the concept of sustainability and the broad concept of health we argue that health and sustainability must be seen as a duality, which could lead towards a better and more equally distributed state of health, globally and locally, and towards more sustainable production and consumption.

The concept of sustainability forwarded by the Brundtland Commission includes health as a part of the social dimension but in practice the question of health remains synonymous with the establishment of a better health care service. WHO has drawn a lot of attention to the increase of lifestyle diseases, and diseases caused by deterioration of the surrounding environment. Also, the more psychosocial consequences of the changed lifestyles have been integrated in their understanding of environmental health. But, in its more practice-orientated initiatives, the focus is primarily directed at better health care services. Thomas et al. conceptualise the problem by introducing the understanding of health in relation to sustainability by making a distinction between 'sustainability *in* health' and 'sustainability *and* health', where the first one refers to health as a production primarily focused on the health care service with its own resources, means of production and economic estimations, while the other definition, sustainability *and* health refers to a far more extensive understanding. Health is here seen as a central focus for all strategies of sustainability with an eye to the achievement of a better health condition globally as well as locally. Health is here understood as a mainstream strategy included in all sectors' strategies of sustainability.

Both the concepts of health and the concept of sustainability have within recent years been reconceptualised. The dominating positivistic approaches have been criticised, and emphasis has been ascribed to the normative and political aspects such as democratisation, quality of everyday life and equity.

Since Brundtland, there have been attempts to objectify the concept of sustainability by expressing level and direction by means of a system of indicators in an attempt to monitor and control the development. The most substantial criticism of the concept of sustainability has come from Wolfgang Sachs. Sachs claims that the attempt to monitor and control development in a more sustainable direction by expressing levels and direction through a system of indicators has failed. 'Sustainable development' has not been detached from the traditional growth-based concept of development. On the contrary, the economic growth-based development has absorbed the concept and made it an instrument for further exploitation of nature. Olsén et al. argue that it is important to develop the concept of sustainability as a normative horizon for orientation, and from this position construct and critically discuss the goals of the economy and production as a scope for unfolding democratic and sustainable everyday lives. Here as we have pointed to, health strategies can become a substantial 'actor'.

The criticism of the concept of health has in many settings been that it has not exceeded the bio-medical understanding of health where the main focus is on the diagnosis and treatment of diseases, and where the key words are causality and evidence. Offered is the broad concept of health where the feeling of coherence and meaning in life, collective care in the daily social context, and societal discourses and responsibility to supporting the individuals is seen as a possible path to improving health conditions globally.

With this reconceptualisation we see an opening appear so health may be seen and discussed in relation to the broad concept of sustainability as described above, and we point at a conceptual duality of thought between health and sustainability. Within such a realm of understanding, societal responsibility is linked with practice and experience of everyday life, which means that the dimension of sustainability can be incorporated in an everyday life practice, and that the health problems are lifted up to a common social responsibility and not just a problem and responsibility for the individual.

Through our case study of food production and consumption we have shown that this correlation must be taken into account both within the environmental, the economic and the social areas. Food production is facing a shifting paradigm where the choices will produce different environmental, health-related, democratic and equality-related consequences. The choice and the journey to the choice, to a high degree, will influence which room will be given to normative discussion of a sustainable and healthy life and to the experience of the citizen. A call for solutions at the societal level is therefore made, that food production considers human health, takes place under sustainable conditions, and that the citizens are given the possibility for participation in decision making.

References

Antonovsky, A. 1987. *Unraveling the Mystery of Stress: How People Manage Stress and Stay Well*. San Francisco: Jossey-Bass Publishers.

Elsass, P. and Lauritsen, P. 2006. *Humanistisk sundhedsforskning*. København: Hans Reitzels Forlag.

Gannik, D.E. 2005. *Social sygdomsteori – et situeret perspektiv*. Frederiksberg: Forlaget Samfundslitteratur.

Giddens, A. 1984. *The Constitution of Society*. London: Polity Press.

Giddens, A. 1991. *Modernity and Self-identity: Self and Society in the Late Modern Age*. Cambridge: Polity.

Halkier, B. 1999. *Miljø til dagligt brug*. Frederiksberg: Forlaget Sociologi.

Ilsøe, D.E. 2006. *Økologisk fødevareforbrug mellem marked, hverdagsliv og visioner om en bæredygtig madkultur*. Ph.D. thesis. Roskilde: Roskilde University.

ISAAA. 2006. *Global Status of Commercialised Biotech/GM Crops. BRIEF no. 35*. [Online]. Available at: http://www.isaaa.org/ [accessed: 9 November 2009].

Jensen, U.J. 2006. Sygdom som erfaring eller fremmedlegeme – sundhed som individuel autonomi eller gensidig omsorg?, in *Sundhed, Udvikling og Læring*, edited by K. Akselsen and B. Koch. Copenhagen: Billesø and Baltzer, 7–29.

Land, B., Pedersen, K.B. and Ilsøe, D.E. 2007. Sundhed og planlægning – sociologiske perspektiver, in *Planlægning i teori og praksis – et tværfagligt perspektiv*, edited by A. Jensen, J. Andersen, O.E. Hansen and K.A. Nielsen. Copenhagen: Roskilde University Press, 364–82.

Lang, T. and Heasman, M. 2006. *Food Wars: The Global Battle for Mouths, Minds and Markets*. London: Earthscan.

McMichael, T. 2002. *On Sustainability, Health and Wellbeing*. [Online]. Available at: http://www.nybioscape.org/bioapproach/McMichael.pdf [accessed: 9 November 2009].

Olsćn, P., Nielsen, B.S. and Nielsen, K.A. 2003. *Demokrati og bæredygtighed – Social fantasi og samfundsmæssig rigdomsproduktion*. Roskilde: Roskilde Universitetsforlag.

Popkin, B.M. 2006. Technology, Transport, Globalization and the Nutrition Transition Food Policy. *Food Policy*, 31, 554–69.

Sachs, W., Loske, R. and Linz, M. 2000. *Greening the North: A Post-industrial Blueprint for Ecology and Equity*. London and New York: Zed Books.

Shiva, V. 1989. *Staying Alive*. London and New Delhi: Zed Books.

Shiva, V. 2006. War Against Nature and the People of the South, in *Critical Perspectives on Globalization*, edited by M.D. Giusta, U.S. Kambhampati and R.H. Wade. Cheltenham: Edward Elgar Publishing.

Thomas, T., Douglas, C. and Cohen, H. 2002. Health and Sustainability. [Online]. Available at: http://www.sustainability.dpc.wa.gov.au/docs/BGPapers/CohenDouglasThomas.pdf [accessed: 1 June 2007].

Towle, N.J. 2004. In Search of Sustainability. *Medical Journal of Australia*, 180(11), 556–7.

Wackerhausen, S. 1994. Et åbent sundhedsbegreb – mellem fundamentalisme og relativisme, in *Sundhedsbegreber – filosofi og praksis*, edited by U.J. Jensen and P. F. Andersen. Aarhus: Philosophia, 49–60.

WHO. 1946. *Constitution of the World Health Organization*. New York, 22 July 1946.

WHO. 2000. *The World Health Report 2000. Health Systems: Improving Performance*. [Online]. Available at: http://www.who.int/whr/2000/en/whr00_en.pdf [accessed: 9 November 2009].

World Commission on Environment and Development. 1987. *Our Common Future*. Oxford: Oxford University Press.

Chapter 13

Linking Mobility, Democracy and Sustainability in an Inclusive Approach to Transport Development in the Global South

Maria Figueroa

There is a quality even meaner than outright ugliness or disorder, and this meaner quality is the dishonest mask of pretended order, achieved by ignoring or suppressing the real order that is struggling to exist and to be served. (Jane Jacobs 1961)

Introduction

Unhindered mobility and widespread accessibility to modern transport infrastructure networks and instantaneous communication technologies are key elements to ensuring participation in today's global economy. Both these elements are far from being available to large segments of the population in dense urban areas in the Global South.[1] Transportation trends in those areas, seen through data and social anecdote, evidence more the unfolding of a daily drama in which conflicts are likely and conventional travel routines exert a high individual and societal cost.

The multiple dis-utilities of travel reinforce pervasive transportation and urban development trends such as: the rapid pace of growth of motorised transportation, the demise of non-motorised options, subsidisation of private motorised travel options at the expense of the public options, scarcity of safe, good quality

---

1 Global South here refers to the entire group of 'less developed countries' that correspond to the group of countries classified by the World Bank as having 'low' and 'middle' income, see more at: <http://siteresources.worldbank.org/DATASTATISTICS/ Resources/CLASS.XLS>. In addition, the present chapter can make a more specific reference to these countries given their low level of motorisation (<100–250 cars/1,000 people), which has gained these countries the name 'less motorised countries'. These countries make up over 5 billion of the world's population, which leaves out only the highly motorised urban centres in Asia and Latin America and the approximately 1 billion who are part of the 'high' income developed world, referred to here as 'industrialised countries or simply as countries in the North'.

alternatives to travelling by car and the individual sense of being locked into a car imperative. In this context, people will buy a car as soon as they can afford one and will give preference to the strongest muscular cars their money can buy. In Asia and the Pacific Region, the increasing motorisation takes place instead through the overwhelming presence of motorised two-wheelers (Zegras and Gakenheimer 2006). This form of increased motorisation has resulted in a displacement of human powered vehicles (bicycles) from the streets and contributed to a high personal risk for pedestrians, all of which in turn increases dependency on motorised modes (Hazards Centre 2002, Tiwari 2002, 2003, Zegras and Gakenheimer 2006).

These trends are enhanced by the existence of a booming unregulated market for second-hand vehicles, and by a type of urban development that creates spatial forms of economic and social polarisation (Harvey 1996). A common pattern is one showing densely packed poor areas and isolation and detachment of middle and upper classes into new and more distant locations. In some extreme cases, the patterns of urban development and travel include the super rich travelling to the city by helicopter and returning home to their new residential enclaves in the fractal cities that characterise the post-metropolis (Soja 2000).

For decades now, the primacy of transport policy in developing countries has been to secure constrained and competing resources to build and enlarge road infrastructure, yet the pace of road building cannot match the pace at which traffic demand grows. As cities continue to grow, new road construction in urban areas needs to be spatially negotiated. Space needs to be gained, often forcefully, from the ubiquitous occupation of urban space by infrastructure, people, non-motorised vehicles, and other many forms of mobile vendors and informal residential, commercial and semi-industrial, semi-permanent constructions; in the process producing a sequel of exclusions and new inclusions. Most analysts have described the resulting form of transport and urban spatial development in developing countries as one that neglects fundamental questions of environmental, social and economic sustainability and one that replicates unequal access and can harbour social instability (Whitelegg 1997, Vasconcellos 2001, Thynell 2003, Low and Gleeson 2003, Zegras and Gakenheimer 2006).

There are no simple solutions to address all these conditions. What is clear is that the foundation of and tools for transport and urban planning that perpetuate current policy and decision making are not attuned to the complexity of urban transport reality in the Global South. The conventional wisdom of planning professionals and politicians still is that transport systems can and will eventually evolve in the right direction and somehow resemble the realities achieved in industrialised countries. The challenges of securing a minimum level and quality of mobility for the vast majority of developing countries' populations are immense and unlike anything industrialised countries have experienced in the evolution of their transportation systems. The approach to achieving transport sustainability in the Global South needs a new and clear reformulation.

The present chapter seeks to advance ideas for the development of a new discursive trajectory that can furnish a different approach. The departing premise is that in developing country cities[2] mobility cannot be disassociated from its political, antagonistic and conflictual elements; transport policy and decision making is inherently an agon that requires a decision, the result of which will always produce losers as well as winners. Finding ways of directly recognising and addressing these conflictual and political aspects of mobility is a key task in the path toward sustainable mobility.[3]

Methodology

The present analysis seeks to provide a platform for understanding the political aspects of transport planning in developing countries and for addressing its conflictual essence. This approach is inspired by theoretical studies on the political as 'a space of power, conflict and antagonisms' (Mouffe 2000),[4] by studies advanced by Vasconcellos (2001) in the case of transport planning in developing countries, by Hillier (2009), Pløger (2008) and Swyngedouw (2007), in the case of urban planning in industrialised countries. These studies have contributed to making the case for restoring the role of politics into planning. The proposition this chapter advances is that to replace the evacuation of politics in transport planning and policy making, a more explicit association between democratic politics and transportation planning for sustainability is necessary. Democratic politics is seen here as the tradition connected to the promotion of a civic culture of equality, association, collective learning, participation and mobilisation as opposed to the more exclusive understanding of the term in relation to a participatory system of decision making.

The chapter's main proposition is that by subjecting transport planning and decision making to a constant political interrogation of its goals, objectives and merits, the conflictual and political aspects can be acknowledged and eventually addressed in developing countries. In particular, the chapter considers what

---

2 Structural factors such as prevalence of weak political systems, weak democratic institutions, low educational levels, high levels of social and economic inequality, poor knowledge of economic and political processes, and weak access to power and decision making common to many developing countries are not contemplated in this analysis. They are considered to make only more pressing the need to carry on the exercise of transport planning with a political 'P' as labelled by Hillier (2009).

3 We use here the concept of sustainable mobility developed by Banister (2008) as an alternative paradigm that includes measures to achieve a reduction of the need to travel, promotion of walking and cycling, and development of a new transport hierarchy of modes, land use development to achieve distance reduction, technology innovation, healthy transport, involvement of the people and public acceptability of measures.

4 In Chantal Mouffe's views, the political is a dimension of antagonism constitutive of human societies, while '"politics" is the set of practices and institutions through which an order is created, organizing human coexistence in the context of conflictuality provided by the political' (Mouffe 2005: 9).

principles would ensure that a minimum level of sustainable mobility (quality and quantity) can be achieved for the majority of urban poor in these countries.

To illustrate the points of the discussion, an example is presented based on a desk review of a number of research papers and official documents on the case of Delhi Metro in India.[5] The review provides insights into the planning and results of Metro Delhi as a public transport infrastructure. The case was chosen as its interpretation can provide a unique wealth of information as a critical and paradigmatic case (Flyvbjerg 2001). The case has strategic importance to both the interrogation of the political that is attempted in this study and to highlight a number of characteristics common to transportation planning in many developing countries.

This study does not include a review of the conditions of governance in developing countries such as the structure of the state, the profile of decision-making processes or the study of the main agents involved. In most developing countries, the public sector is formally in charge of a fragmented set of agencies dealing with urban planning and transport decision making (Vasconcellos 2001). The chapter's attention is primarily placed upon the character of the project at hand, the expressed objectives in its planning as a public transport service and the goals initially achieved upon implementation. This approach is then generalised under the political interrogation of what could be an alternative narrative to framing the problem departing from how democratic principles inspire a sustainability approach to the provision of public transport services. This discussion has relevance to governments in the Global South that operate, at least, under a veil of participatory democracy.

Approaching the Political in Mobility

The task of spatial planning involves the distribution of power and resources and is influenced by party politics. Consequently, planning is inherently political (Hillier 2009: 1). The assumptions and the role of planning and spatial decision making need to be subjected to 'a political interrogation through which relations between space, politics and contestation to management become apparent' (Ibid: 1). To Hillier the important question to keep asking is what does planning do?

The literature addressing the political in transport planning in general and in the case of developing countries in particular is extremely limited. One exemption is the seminal work of Vasconcellos (2001) who documented how the physical arrangement of most large cities in developing countries are 'proof of the shaping of circulation space for the most powerful roles, especially the driver, more directly the middle-class-driver' (Ibid: 75). Vasconcellos's work can be seen as a

---

5   The motivation to use this case was provided by direct interaction with the researchers and the work carried out by the Hazards Centre (2006) and the Transportation Research and Injury Prevention Program (TRIPP) of the India Institute of Technology on the Delhi Metro.

precursor in advocating the need to establish a political interrogation of transport planning decisions. The approach he suggested departs from understanding the different roles played by people using transport as pedestrians, passengers in cars, drivers, passengers in public transport, residents, workers, waiting for the bus, and the conflicts they engender of their daily interaction which to him encapsulate a political content.

From a sociology of mobility perspective, Sheller has suggested that one of the effects of urban mobility on societies is that it creates 'a technoscape that ... performatively produces particular subjetivities such as "the driver, the pedestrian, the biker, etc."' (Sheller and Urry 2006: 9). Taking this proposition forward, Richardson and Jensen suggest that in fact a new form of 'idealised mobile subjects' has come to be pre-designed or fabricated by planners: 'a transit system may form part of a complex strategy of managing mobility ... a project move from ideas to drawing board to reality, in which policy makers, planners, architects and urban designers construct idealized mobile subjects that will inhabit them' (Richardson and Jensen 2008: 226). Even though their approach is sociological and post-political, I will use their description of 'pre-designed idealised mobile subjects' to help explain how political inequality is reproduced spatially in connection with the expected transit riders of Metro Dehli.

Vasconcellos's (2001) political approach to transport planning is constructed as follows: people enter into conflict with the 'transport' roles played by others; and such conflicts can be only partially moderated by the traffic rules and by individual minor decisions as traffic planning and management only deals with the physical characteristics of the conflicts. Therefore, planning cannot be a neutral activity but a political one, in which a negotiation is always necessary (Vasconcellas 2001: 78). For the purposes of this chapter, a similar basic distinction is made to identify transport planning and decision making as an agon in which the necessary decisions will have losers as well as winners.

Four key questions are essential to Vasconcellos's (2001: 57) political interrogation of transport planning and project decision making: (1) Who is behind the decision and in charge of planning the project (elite, financial sector, industry)? (2) What are the symbolic characteristics attributed to this project (i.e.: progress, clean urban space, etc.)? (3) How is the project ruled and operated after construction (i.e.: connected or disconnected to the transport network serving the poor)? (4) Whose role is favoured with this transport planning intervention (who are the users of the new system but also who is obtaining benefits from land use conversions, etc.)? For our purposes all these questions are relevant to help explain part of what constitutes a significant and complex set of forces including, for example, how different travel demands from people belonging to different income groups are served, what forms of spatial and modal exclusions are made to accommodate the newly accepted inclusions, which actors and coalitions of actors benefit, what other hidden costs to society are not being made apparent limiting the use of scarce resources in other areas.

A different type of analysis of the political is proposed by Hillier (2009) who constructs her interrogation based on Deleuze and Guattari's cartography of tracing which departs from asking the question 'how did something come to be?' The result of Hillier's analysis is a topological investigation similar to a 'Foucaldian genealogy paying special attention to the forms of strife or competition between different ways of connecting, controlling and framing issues' (Hillier 2009: 5).

The analytical structure this chapter puts forward can be situated partly between these two approaches. In its essence, it builds strongly on the set of questions about the political character of the transport planning interventions of Vasconcellos, but it complements these questions by establishing connections between them and pre-determined and well accepted democratic principles seeking to problematise the existing planning narrative that justifies the transport projects as one that doesn't even attend to a minimum set of democratic principles. The attention to a pre-defined set of principles make this analysis different from the topological cartographic developed by Hillier (2009).

The present approach seeks to merge the sustainable mobility debate and planning practices as part of the essential discourse of democracy, recapturing issues of individual and social responsibility, materialisation of policies that privatise space and limit public access and mobility as losses a democratic society cannot afford to accept. The resulting discussion allows some speculative proposals on the 'what if' of the relations-to-come and about what is the significance of these speculative relations for creating an alternative transport planning narrative where goals such as a minimum quality of sustainable mobility services for the poorest and voiceless of the world can be contemplated.

One key question in the present analysis will be then 'what if?'; in the sense of what would be the implication of changing the terms of definition of the problems and the solutions in transport planning, so that it seeks to achieve a democratisation of a minimum quality and quantity of mobility services in developing countries? The attempted reflections will also ponder about the kind of balance toward sustainability this type of realisation may entail under the conflictual circumstances of the example case considered.

Democratic Principles and the Mobility Debate

Democratic principles, norms and the theoretical foundations of democratic thinking are frequently used to evaluate decision-making processes and issues of legitimacy in planning (Figueroa 2006, Wiklund 2002). The present chapter seeks to move beyond the question of democratic legitimacy to relate democratic principles to a mobility discussion as the basis for a political interrogation of transport planning practices and seeking a new narrative and new thinking about possibilities concerning how to achieve the conflictual goals of transport sustainability in the Global South.

Democratic principles differ according to two basic traditions: liberal and communitarian. The liberal tradition conceives politics as private in nature and instrumental in purpose (Machepherson 1977, Wiklund 2002: 54). Politics is characterised as a market where actors (consumers) express their private self-interested and pre-determined preferences. Political participation is purely instrumental and actors act strategically to achieve their goals. Democracy under this tradition becomes a mechanism for the 'fair and efficient' aggregation of political actors' private interests into collective choices (Wiklund 2002).

The communitarian or republican tradition characterises politics as public in nature and non-instrumental in purpose (Wiklund 2002: 55). Politics is not seen as a market but as a forum, political actors are not strategic consumers but active participants. Through such participation they can orient themselves towards the common interest, become aware of their dependency on one another, and learn to recognise the value of their participation which resides in the potential to educate and transform political actors and their interests (Pateman 1970). Politics under this tradition is viewed as an end in itself (Elster 1997: 34, Held 1996: 36). Democracy is more than the simple process of aggregation of private opinions and interests, becoming an integrative force for the members of a society, which are bounded through discussions, the discovery of their common interest and their shared traditions and by their ability to affect the decisions that will affect them. Departing from these two traditions a great number of conceptual interpretations and forms of exercising the democratic process are possible.

Considering the two most fundamental distinctions in the core notion of democracy is necessary before proceeding on the discussion on the links between the democratic principles, mobility and sustainability. In the proceeding analysis, however, the chapter will minimise the focus on the distinction between democratic principles according to the libertarian and communitarian traditions. This position is based on the assumption that within any democratic society a positive development in the direction of the goals of the integrative tradition (i.e.: greater participation, transformation of political actors from self-interested motivations into awareness of their dependency and common interest, etc.) will have a positive impact on the goals of the aggregative tradition goals (i.e. a fair and efficient aggregation of self-interested opinions). Conversely, a closer inspection into the bundle of elements that contribute to a 'fair' and 'efficient' aggregation of interest, can only be facilitated by a better informed, educated, and with a good sense of the common good, set of individuals in a society. Thus the chapter will continue to make references to democratic principles without any further reference to these two traditions.

The following discussion has been organised into two major focus areas defining linkages between one or two democratic principles and mobility. The two areas have a strong explanatory power on matters of relevance to the political in a developing country context, to transport sustainability in general and to the example discussed in particular.

I) The Principles of 'Equality and of Individual Liberties'

Whether self-interested or community-oriented, a democratic society consists of the *demos*: in essence the people, the notion that all persons can exercise a right to voice/vote their opinion and that it will count equally in the larger society's decision-making processes. In today's modern world there is overall agreement and human aspiration that a minimum of conditions directly affecting the quality of life of individuals should be democratised, for example education and health care. In the same manner, mobility has arguably become a form of democratic global aspiration. The formulation of a democratic mobility principle could even be of value. One possible wording for this principle could read as: the equal right of individuals to achieve a certain level of mobility that would improve the conditions of their lives facilitating the fulfilment of their familial, social and cultural obligations and the exercise of their individual civil, political and economic liberties.

While at the society level guaranteeing a minimum level and quality of mobility for all can have measurable benefits, at an individual level, if everyone could maximise their individual liberty to mobility, following the way we conceptualise mobility today, will become inherently destructive as there is a capacity threshold beyond which additional travel hinders every other traveller's mobility and decreases their utility.

The goal of equality in transport planning cannot be understood then as an equal, unlimited right to mobility. It needs to be pursued instead through a demonstration of awareness of the type of inclusions and exclusions that every decision entails. For example, in terms of the poor's access to the planned transport services, allocation of investments between modes, allocation of rights over the use of roads, safety management of non-motorised modes. One of these aspects, disabled people's equality of access to public forms of transportation has received increasing attention in recent years (Rickert 2005). Equality as a principle includes turning the focus in terms of research and planning effort to understanding and creating opportunities for equal access and mobility for women, children and the elderly which in developing countries are palpably under-represented.

Among industrialised countries, nations like the Netherlands, Denmark and Sweden today demonstrate one of the most advanced forms of a functional egalitarian transport system. Unsurprisingly, these same countries are identified by analysts as leaders in transport sustainability implementation (Low and Gleeson 2003, Nilsson and Neergaard 2009). The more individualistic car-oriented systems found in countries like the United States and Australia at the other end of the spectrum exemplify how advocating non-egalitarian mobility can only lead to unsustainable forms of transport development.

The question remains whether these results are related to other socio-cultural characteristics such as the fact that in Europe the populace in the various countries appears to feel enough sense of social connectedness, where the enforcement of a social contract that benefits all such as health care, welfare or a comprehensive public transport system can be more readily accepted even if at a fairly high cost. In the United States and other similar countries, the society as a whole relies more on unfettered individualism than on social solidarity, making it more difficult to contemplate a transition toward a more egalitarian pattern of mobility. All in all, these are the extreme cases of urban and transport development that developing countries so far have sought to emulate.

Another potential way of framing this debate in a form relevant for developing countries is by asking: Can mobility be related to individuals' 'needs' as defined within the sustainability debate? Should the goal of sustainable transport planning be to guarantee the 'travel needs' of the present as well as future generations? The association of travel with a 'needs' discourse suggests that there is something deeper at work than a utility that is gained from travelling. Access to travel and mobility become an intrinsic necessity of social existence in an urban area. Access to a minimum quality and level of mobility thus can be seen as a basic condition in support of the urban spatial condition of human development in a global context.

Could the goal of democratising a minimum level of qualitative and quantitative mobility serve as inspiration to transportation planning decisions in the Global South? At a minimum, can individual projects be evaluated according to how much they contribute toward a goal of providing mobility to the disenfranchised non-motorised many? The example of the Metro Delhi will further reflect on the way transport planning in countries like India can or are willing to address this question.

II) The Principles of Shared Responsibility, Interdependence and the Common Good

Transport planning has as its basis a libertarian approach to understanding travel demand. Individuals have rights to exercise their preferred amount and ways of travel and the planner's job is to understand and aggregate those preferences (travel demand) and to come up with solutions that will facilitate the exercise of each individual self-interested form of mobility.

Travel and mobility decisions are in reality not intrinsically individual. Each physically travelling individual partakes within the same limited spatial configuration. Therefore, something akin to mutual self-interest needs to be found in this context (Whitelegg 1997). Moving around individually affects the whole; the movement of the whole affects the individual through increasing congestion and pollution, for example.

The compartmentalisation in planning of mobility decisions between public and private and the effort to prioritise services that improve primarily private motorised mobility has reduced transport planners' principal task to one of finding ways for accommodating the high-speed movement and spatial parking demands of private motor vehicles. This goal is spatially and financially unattainable. Transport projects and policy decisions that benefit collective movement should have an intrinsic greater value to society than those that benefit private movement. Transport policy that subsidises and encourages increasing private motorised transport to facilitate better mobility disregards the fact that individual mobility affects the common and the condition of shared responsibility that should exist in the use of an intrinsically limited urban space. The critical questions of what is transport planning to do, and to whose benefit are transportation interventions planned, become urgent as the Metro Delhi case will further help exemplify.

The Metro Delhi Case

> It will be much more than a cheap and safer means of transport. It will reduce congestion on roads making movement easier. It will also reduce atmospheric pollution to a great level making the environment healthy... The Metro will totally transform our social culture giving us a sense of discipline, cleanliness and enhance multifold development of this cosmopolitan city. Delhi Metro Managing Director, E. Sreedharan, from an article in The Hindu by Sandeep Joshi. (quoted in Siemiatyck 2006: 281)

Delhi exemplifies a metropolis where the complexity of the challenges posed by the transportation sector is by all accounts extreme. The quote above reflects how Metro Delhi gave city planners an opportunity to engage in the cultivation of creating a good image for the city. Nowhere, however, would the main media have reported a statement like the one above contrasted by another like '... the development of Metro Delhi has been accompanied by land appropriations, slum clearances, and a broad range of political and economic opportunism that threatens to exacerbate the rift between the wealthy and the poor' (Siemiatyck 2006: 278). The contraposition of the realities that make the utterance of these two statements possible is what a focus on the political in planning can help identify and eventually address.

Delhi is the capital of India, the second most populated country in the world. Its population currently exceeds 13 million and is expected to increase to 23 million by 2021 (Delhi Development Authority 2003). The sheer size of the population and the low per capita average income around US $1,190[6] (Economy Watch 2005) gives an idea of the scale of passenger travel activity that needs to be channelled by means of affordable public transportation services. The problem of affordability gives new meaning to the concept of what constitutes a public means of transport in Delhi and is one deterrent for the realisation of a huge potential market of public transport riders in conventional means, buses, trams, metro. In fact, many city residents are so poor that they cannot afford even low transportation fares, they travel on foot, bicycles and non-motorised rickshaws to meet their mobility and economic needs, i.e.: mobile vendors (Tiwari 2002, 2003, Polakit 2008).

The opportunities that a minimum level of mobility services could bring in terms of access to services and jobs are simply not easily delivered to the poor

---

6   Comparatively the annual average per capita income of a poor family in the United States is US $23,400 (Cotter, England and Hermsen 2007).

at conventional market price levels. A rising middle class, on the other hand, and the combined effect of media and the steady progression of Western values have generated a cosmopolitan and far looking society. Banking on these sentiments, the public authorities have been eager to demonstrate how India is prepared to be integrated into the global economy – an opportunity that the development of Delhi Metro, amongst other things, has made possible.

The mobility and accessibility options of the poor in Delhi are restricted, of poor quality and mostly informal, a situation that in many respects is typical in other developing countries. However, there are some particular aspects to the traffic situation in Delhi: '… the situation on any given road in Delhi including major highways is one where cars, trucks, buses, motorcycles, and mopeds – or a myriad of motorized modes compete for space with three wheeled auto-rickshaws, horse-drawn carriages, donkey-drawn wagons, human-pulled carts and pedestrians, resulting in extreme congestion, road accidents, air and noise pollution' (Siemiatycki 2006: 279).

Traffic in Delhi has been predominantly dependent on road transport, with railways catering to only about one per cent of the local traffic till 2003 (Tiwari 2002, Economic Survey of Delhi 2005). The complicated congested state of the roads, in addition to the disperse, poly-nuclear land use pattern of development, recently featuring a number of new satellite towns on the outskirts of Delhi, have contributed to the exponential growth of private cars during the last decades. Private motorised vehicles are popular as they represent both a symbol of status and of freedom from deteriorating conditions of public transport.

The dramatic increase in motor vehicles has only added to the congestion problems creating a negative feedback loop. In 2003, buses constituted about one per cent of the total number of vehicles, while they served 60 per cent of the total traffic demand (Economic Survey of Delhi 2005: 131). Public buses provide a low level of service and comfort, with passengers often travelling on footboards. Large-scale privatisation increased the capacity but buses continue to be overcrowded and poorly maintained. Buses receive no preferential treatment in terms of dedicated lanes or traffic management, despite carrying more than half of all passenger travel in the city (Tiwari 2002).

Delhi Metro was planned as a technology exchange with a number of international sponsors and the goal of producing a drastic overhaul to Delhi's transportation landscape. The emphasis was given to contracting out tasks internationally while creating partnerships with Indian firms to ensure the transference of expertise. The financing of the project was obtained through a combination of international and local funding sources: 30 per cent was financed through equity contributions subscribed equally by the Central Government and the Delhi Government; an interest-free loan covered the cost of land acquisition equivalent to eight per cent; another 56 per cent financed by the Japanese Government; and finally the last six per cent was planned to be met by raising money through property development[7] (Roy et al. 2006).

---

7   <http://www.dmrc.delhigov.in> accessed 15 September 2009.

The company, Delhi Metro Rail Transport Corporation, was created in 1995 with an equal partnership between the national and city government of Delhi. The company was commissioned with building the two phases of the Metro, a 65.10 km route in Phase I and is proceeding with another 121 km in Phase II.[8] The new company represented a departure in the management of urban transport from the common situation in India where transport has not been an area of systematic concern and where planning institutions in general are still evolving. The Ministry of Urban Development is today the nodal agency for transport policy and planning and it only recently formulated its first National Urban Transport Policy in 2006. Neither the state nor the local level have the capacity for holistic urban transport planning and there is a high level of fragmentation, piecemeal planning and implementation, coupled with poor coordination.[9]

In this light, the newly founded Metro Corporation has been able to preserve a mantra of efficiency and good management, avoiding accusations of bribery and corruption that often plague major infrastructure projects in India (Siemiatycki 2006: 288). A number of studies, however, have pointed at the way in which opportunism and the influence of a number of coalitions has permeated the development of the Delhi Metro (Siemiatycki 2006: 288). Coalitions like this have been identified in a study of Asian Metro developments to include politicians, property owners, planners, business groups and others who together or separately have been capable of exercising pressure to guide decisions in their favour (Townsend 2003).

The main objective of Metro Delhi was the development of a mass rapid transit system with the Metro as its backbone. A number of other goals transcending the simple movement of people have accompanied the planning, design and implementation of this project. The Metro was promoted for its potential to provide: reduced travel times, increased comfort and improved safety – especially for women – and, as a symbol of environmental care to the public. Beside those, Delhi Metro's promotional message attributed all aspects of its development to the creation of a vital piece of infrastructure that would help transform the city into a modern metropolis (Siemiatycki 2006: 283).

Through the planning and development of the Metro, Delhi planners were able to bring hope to the idea that technology progress can alleviate the challenges the city of Delhi with its mounting population faces to provide urban mobility, sanitary conditions, employment and security (Siemiatycki 2006: 285). The attention given to design features mixing ultra-modern technology with visually striking designs and features to make the system accessible to disabled people are still presented as examples that Delhi has become ready to compete as a world class city.

---

8   <http://www.delhimetrorail.com/corporates/needfor_mrts.html>   accessed   15 September 2009.

9   This description is based on power point presentation notes by Awargal, O.P. (2008): *Planning and Decision Making Institutions and Approaches for Urban Transport in India.* Ex-Head of Urban Transport Division, Ministry of Urban Development, India. Presentation given at EMBARQ, World Resources Institute.

Delhi Metro has thus been a vehicle for inculcating a culture of discipline, order, routine and cleanliness as a vision for a modern city that the planners expect to extend beyond the station walls (Siemiatycki 2006: 285). In this process, however, 'industrial units have been closed down, slums have being removed, there are investments in large infrastructure, utilities are being privatized, and older systems of governance are being steadily eroded, all to create a political perception of law and order, of disciplined citizens, and of cleanliness capable of attracting further foreign investments and activities' (Randhwava et al. 2006: 12). Finally, a well-articulated local and international media effort making favourable comparisons of the Delhi Metro to other urban systems around the world has also allowed the promotion of a bright future for Delhi as a high-tech, rationally planned, competitive city and it has assured a widespread public support for the project locally.

Against this background, how is this new infrastructure meeting the transportation needs of the larger community it intends to serve? The following discussion looks into three indicators of results achieved so far by Delhi Metro in terms of level of ridership, and the exclusions and inclusions the planned redevelopments have produced, contrasting these results with the principles of democracy described in the analytical framework to help answer this question.

Antagonisms, Hope and Political Hubris: The Faces of Global South Mobility

The conditions of urban transport planning defined in the previous section can be said to exist even in developing country cities with a long tradition of financial investment in transportation infrastructure. Transport planners and decision makers are, in general, given considerably more power to shape the city with the development of large infrastructure projects, such as Metro. The idea that these investments will spur economic growth and prosperity is a powerful one and it remains unclear how much are, governments in developing countries and elsewhere, interested in pulling back from any opportunity to induce a fast-growth development model, regardless of the consequences.

In the presence of super investments in transport that are run by especially appointed companies, the regulatory power of local authorities fades and what may have otherwise served to create some oversight controls from citizens (Hazards Centre 2002, Roy et al. 2006) or the local government tend to be released in order to secure the expected potential economic growth. This gives place to a form of transport policy hubris which in practice and paradoxically ends up compromising the ability of the costly infrastructure to meet the transportation needs of the community it intends to serve.

One of the consequences when planners act under these conditions is that the process becomes one of designing who the new 'idealised mobile subjects' as described by Richardson and Jensen (2008) will be. The design of idealised subjects can be seen as giving opportunity to a subtle process of exclusions embedded in the planning of a big transport infrastructure like a Metro. In Delhi Metro, this is exemplified by the emphasis awarded to a new aesthetics in the stations and surrounding the stations that included 'an open concept layout, technologically advanced no-touch turnstiles, security cameras and well-appointed station trimming projects, silver trains with a sleek design, automatic doors, digital signs and climate control all enacting an image of progress, order, cleanliness and security which contrast sharply with the congested, unpredictable chaotic pace of life in Delhi' (Roy et al. 2006). Another example is the number of rules of conduct for passengers inside the Metro which extend a tight control over the riders going well beyond general measures of safety: 'the metro operators have been vested with a wide-range of powers to maintain security with a firm stance on who has legitimate entitlement to access the promises in the system'[10] (Siemiatycki 2006: 285).

A number of studies have reported the fact that the Delhi Metro caters mostly to a small section of more affluent private riders who can afford the fare structure, and travel the distance to the Metro station (Siemiatycky 2006, Roy et al. 2006). The pattern of redevelopment of property on both sides of Delhi Metro lines indicates an uneven distributional impact of the neighborhood revitalisation efforts. The result is as Siemiatycky describes it that while 'the educated, the wealthy and powerful are invited to turn their gaze to the world, ... the poor are seeing their homes disappear for a development they do not have the skills or income to benefit from' (Siemiatycky 2006: 287).

Ridership on Delhi Metro appears to be well below even revised expectations (on average 250,000 daily riders or 17 per cent of the revised ridership forecast for 2005 (Sharma 2005, cited in Siemiatycki 2006: 286). As the ultimate goal of providing better transport options to the Delhi commuter vanes, it becomes clear that Delhi Metro has encouraged real estate business in the city in what amounts in reality to a 'huge transfer of public money into private pockets and a distribution of social and environmental costs over a much larger population that will not even travel by Metro' (Randhwava et al. 2006). Clearly the creation of a 'world class' city through the development of large and costly transport infrastructure is in dire contrast with the possibilities of creating a democratic city with opportunities of access and mobility for all.

Conclusions

Can developing countries create a 'different' sustainability narrative form of transportation development? To answer this question, planners need to define better tools to address a number of difficult political challenges such as social exclusion, informality of transport services and forms of employment, social and environmental justice, and the co-existence of culturally diverse forms of providing

---

10   Terrorism has increasingly become a real (if not worse) problem for the authorities to sort out. During recent bombings in Delhi, thanks to the high security and the metal detectors, the bombs went off above the metro and not in it.

mobility services. Democratically committed planning practices can begin to create opportunities for emergence of a new narrative and for a re-conceptualisation of research and planning tools that can set the tone for an open discussion and assignation of responsibilities from the individual to the larger society.

Transport planning in the Global South cannot continue to be practised as political hubris. The terms and basic premises, knowledge tools and pool of commonly accepted projects and decisions need to become the subject of political interrogations, open public scrutiny and reconsideration. The expectation that a steady progression from chaos to order will unfold, with one costly project investment at a time, only results in exclusions, privatisation of space and negation of a fundamental and ethical right to the city to most of its citizens, as the Delhi Metro exemplifies.

The general picture of transportation in the Global South of congestion, pollution, noise and chaos needs to be revisited not as a form of extenuating circumstance in a turbulent path to a modern transport system, but thoroughly researched, acknowledged and comprehended in its own political dimension, including accepting the fact that a full reconciliation with the modern individualistic mobility paradigm cannot be expected. There is no possibility of a smooth evolution here. An inclusive approach to sustainable transport development will need to accept that informal aspects of mobility may be an asset toward affordable solutions to the poor and that it needs to develop new tools for integrating them as such.

A new sustainability agenda for transport development in the South could emerge from this effort to discuss and give shape to a new range of creative ideas from the chaos to sustainability, maybe never directly touching upon the order found in Western modern societies. Research is fundamental to help highlight this new focus and sense of direction, also to highlight primarily the difficult trade-offs. A redefinition of academic inquiry becomes necessary, with attention to the fact that as problems and priorities are different so too are the range of solutions.

The association between mobility, its political dimensions and democratic practices of mobility here discussed is an initial attempt to force a new narrative; one that clearly is still in need of further elaboration. This contribution shows, however, that linking mobility and democratic thinking could have powerful implications to clarify the tasks at hand regarding sustainable transportation. It makes at the least very clear that greater education, employment, equality of participation, sense of shared responsibility and provision of a minimum quality of democratically accessible mobility services are all important components of the implementation of sustainable mobility solutions in the Global South.

References

Banister, D. 2008. The Sustainable Mobility Paradigm. *Transport Policy*, 15, 73–80.
Cotter, D., England, P. and Hermsen, J. 2007. *Moms and Jobs: Trends in Mothers' Employment and Which Mothers Stay Home. A Fact Sheet from*

Council on Contemporary Families. New York: Department of Sociology, Union College, Schenectady.

Delhi Development Authority. 2003. *Delhi Master Plan.* [Online]. Available at: http://www.dda.org.in/planning/mpd-2001.htm [accessed: 14 October 2009].

Economic Survey of Delhi. 2005. *Transport.* Delhi: India's Department of Planning, 130–46.

Economy Watch. 2005. *Per Capita Income: Delhi No. 2.* Economy Watch. [Online]. Available at: http://www.economywatch.com/indianeconomy/poverty-in-india.html [accessed: 14 October 2009].

Elster, J. 1997. The Market and the Forum: Three Varieties of Political Theory, in *Deliberative Democracy: Essays on Reason and Politics*, edited by J. Bohman and W. Rehg. Cambridge: MIT Press.

Figueroa, M. 2006. *Democracy and Environmental Integration in Decision-Making: An Evaluation of Decisions for Large Infrastructure Projects.* Ph.D. Thesis. Roskilde: Roskilde University.

Flyvbjerg, B. 2001. *Making Social Science Matter: Why Social Inquiry Fails and How it can Succeed Again.* Cambridge: Cambridge University Press.

Flyvbjerg, B., Bruzelius, N. and Tothengatter, W. 2003. *Megaprojects and Risk. An Anatomy of Ambition.* Cambridge: Cambridge University Press.

Harvey, D. 1996. *Justice, Nature and the Geography of Difference.* Oxford: Blackwell.

Hazards Centre. 2006. *Restructuring and Efficiency The Case of Delhi Transport Corporation, Views from Below.* New Delhi: Sanchal Foundation.

Held, D. 1996. *Models of Democracy.* Cambridge: Polity Press.

Hillier, J. 2007. *Stretching Beyond the Horizon: A Multiplanar Theory of Spatial Planning and Governance.* Aldershot: Ashgate.

Hillier, J. 2009. *Planning with a Political 'P?'* Paper to Workshop on Planning, Development and Resistance. ETH Zurich, 12–13.

Hirschman, A. 1994. Social Conflicts as Pillars of Democratic Market Society. *Political Theory*, 22(2), 203–18.

Jacobs, J. 1961. *The Death and Life of Great American Cities.* New York: Random House.

Low, N. and Gleeson, B. (eds) 2003. *Making Urban Transport Sustainable.* New York: Palgrave Macmillan.

Macpherson, C. 1977. *Life and Times of Liberal Democracy.* Oxford: Oxford University Press.

Mouffe, C. 2005. *On The Political: Thinking in Action.* New York: Routledge.

Nilsson, A. and Neergaard, K. 2009. Åtgärdsanalys av EU:s transportpolitik. Rapport 2008:93. Lund: Trivector Traffic AB. [in Swedish].

Pateman, C. 1970. *Participation and Democratic Theory.* Cambridge: Cambridge University Press.

Pløger, J. 2008. Foucault's Dispositive and the City. *Planning Theory*, 7(1), 51–70.

Polakit, K. and Boontharm, D. 2008. Mobile Vendors: Persistence of Local Culture in the Changing Global Economy of Bangkok, in *Local Sustainable Urban*

Development in a Globalized World, edited by L.C. Herbele and S.M. Opp. Aldershot: Ashgate, 175–200.

Rickert, T. 2005. *Transport for All, What Should We Measure.* [Online] Access Exchange International. Available at: http://www.globalride-sf.org [accessed: 9 November 2009].

Richardson, T. and Jensen, O.B. 2008. How Mobility Systems Produce Inequality: Making Mobile Subject Types on the Bangkok Sky Train. *Built Environment*, 34(2), 218–31.

Roy, D., Randhawa, P., Goswami, M. and Barua, B. 2006. *Delhi Metro Rail: A New Form of Public Transport?* New Delhi: Hazards Centre, A Unit of Sanchal Foundation.

Siemiatycki, M. 2006. Message in a Metro: Building Urban Rail Infrastructure and Image in Delhi, India. *International Journal of Urban and Regional Research*, 30(2), 277–92.

Sheller, M. and Urry, J. (eds) 2006. *Mobile Technologies of the City.* London: Routledge.

Soya, E. 2000. *Postmetropolis: Critical Studies of Cities and Regions.* Oxford: Blackwell.

Swyngedouw, E. 2007. The Post-political City, in *The City of Late Capitalism and its Discontents*, edited by BAVO (Gideon Boie and Matthias Pauwels). Rotterdam: NAI-Publishers.

Thynell, M. 2003. *The Unmanageable Modernity: An Explorative Study of Motorized Mobility in Development.* Ph.D. Thesis, Department of Peace and Development Research. Göteborg: Vasastadens Bokbinderi.

Tiwari, G. 2002. Urban Transport Priorities Meeting the Challenge of Socio Economic Diversity in Cities – Case Study Delhi India. Transportation Research and Injury Prevention Program (TRIPP). New Delhi: India Institute of Technology.

Tiwari, G. 2003. Encroachers or Service Providers. Transportation Research and Injury Prevention Program (TRIPP). New Delhi: India Institute of Technology.

Townsend, C. 2003. *In Whose Interest.* Ph.D. dissertation. Perth: Murdoch University.

Vasconcellos, E. 2001. *Urban Transport, Environment and Equity: The Case for Developing Countries.* London: Earthscan.

Wiklund, H. 2002. *Arenas for Democratic Deliberation: Decision-Making in an Infrastructure Project in Sweden.* Jönköping: International Business School Dissertation Series, No. 013.

Whitelegg, J. 1997. *Critical Mass. Transport, Environment and Society in the Twenty-first Century.* London: Pluto Press.

Zegras, C. and Gakenheimer, R. 2006. Driving Forces in Developing Cities' Transportation Systems: Insights from Selected Cases. Working Paper. Massachusetts: Massachusetts Institute of Technology.

Conclusions and Perspectives for Research and Political Practice

Bo Elling, Maria Figueroa, Erling Jelsøe and Kurt Aagaard Nielsen

This chapter is divided into two parts: the first, following the outline of the book, summarises the contributions to the new sustainability agenda proffered by the different authors. The second part concerns themes common to the chapters, which suggest future research in the area, namely health, interdisciplinarity, social responsibility, and participatory and dialogical democratic initiatives.

Concluding Summary

Part I: Philosophical, Ethical and Meta-theoretical Aspects

The chapters in this section suggest that the rhythms of life, spaces for dialogue and reflection, global ethics and the social framing of systems-based concepts for sustainability are essential dimensions for new theoretical inputs to an agenda for sustainability.

Sustainability requires thoughtfulness and reflection, however sustainability is created in our practices. Such practical forms are characterised as rhythms in Hvid's analysis. Moreover he demands that such practical rhythms of social, economic, natural, environmental and specific everyday conditions of life must be acknowledged when bringing about the changes that are necessary for sustainability. A theoretical renewal toward a holistic conceptualisation of sustainability must encompass such rhythms as a basis for understanding its dynamics.

Practical rhythms and sustainable lifestyles must be the result of negotiations that include ethical claims that should not be overly dominated by either the market or the political administration, or by citizens' interests alone argued by Elling. We face today a kind of asymmetrical realm in which standards can vary considerably depending on who is involved. Creating democratic spaces for such dialogues and reflections can lead us to new ways of acting and thinking that could conceivably 'lead the way' out of contemporary domination by specific interests, and beyond the domination of market forces over political initiatives.

The analysis of several hundred years of sustainable practice in agricultural behaviour on the Faeroe Islands presented by Brandt is historical more than theoretical. But its theoretical implications are nevertheless unavoidable since Brandt highlights the contradictions between ecological and social sustainability and explains how a social framing of sustainable development can benefit these different system-based concepts of sustainability.

Finally, Agger and Jelsøe demonstrate the diversity of positions held in the Danish Council of Ethics regarding the use of utility as a criterion for assessing the release of genetically modified organisms into the environment, despite agreement that sustainability should be part of the normative basis for a decision. They conclude that there are markedly different understandings of the normative character of risk assessment among the members, and they find that for the scientists it is important to be explicit about the foundational values of their views. Sustainable development can be a useful conceptual framework for understanding the complexity of the global society and its relationships with nature, and through the discussion of the GMO case they underline the significance of the ethical issues related to sustainable development.

Part II: Planning and Transition Towards Sustainability

New concepts and approaches to sustainable developments are essential but if they cannot be translated into new planning initiatives and practical experiences, theoretical renewal becomes fruitless. The contributions in Part II give evidence to the challenges that planning, transition management and regulatory approaches have to further practical implementation of sustainability in different areas of intervention: economy, agriculture, health, chemical regulation. The value of the variety of experiences presented in this section is that they provide new conceptual approaches and further insights towards a new sustainability agenda.

Hansen, Søndergård and Stærdahl conclude that a number of changes in transition management are necessary in order to make room for radical innovations towards sustainable production. The first is to ensure variation of transition arenas and accept conflicts. The second is to take the politics of transitions seriously and get support from the general public to outbalance incumbent powerful actors. And the third is to differentiate the governance structures of transition processes.

REACH is often presented as a major step forward in ensuring a more sustainable production and risk-free handling of chemicals. But according to Rank, Syberg and Carlsen many loopholes have to be filled before it can be claimed to contribute to a proper sustainable development. First, they point out that all chemicals must be regulated. Second, they call for more clarity, in particular as regards the authorisation system. And third, they state that a valid risk assessment of all chemicals is not part of REACH. The overall conclusion is that even though REACH is a step forward, it is far from being a sustainable chemicals policy.

In the agriculture field sustainable development is most frequently addressed as market-led environmental change instead of planning steps. Jelsøe and Kjærgård show that this is despite the fact that social dimensions are formally recognised. The limitations in promoting and ensuring changes towards a more sustainable society have much to do with a restrictive conception of sustainability, and must be seen as a significant challenge to the formulation of sustainability policies. Finally, they argue for introducing the concept of food democracy if more sustainable practices are to be achieved and the need to establish pressure for new forms of social regulation inside the food sector.

Crabtree shows that the Brundtland approach to basic needs raises a number of unresolved questions and that, as a consequence, the Commission's definition of sustainability is fundamentally flawed. As an alternative he presents the concept of capabilities, originally developed by the Indian economist and philosopher Amartya Sen. The capabilities approach emphasises the substantive opportunities people have to lead lives they value.

Hansen suggests a number of adaptations in the economic standard model for balancing environmental and economic objectives. Used as a point of departure for economic analysis, these adaptations will make the model more useful in sustainability analysis. However, such adaptations imply a more complicated image of societal deliberation necessary to reach conclusions about environmental-economic analysis. At the same time it provides a framework much better suited for interdisciplinary analysis.

Part III: Life-politics, Democracy and Sustainability

In Part III a broader perspective on sustainability is given though still at the practical level. The perspective is widened not only by expanding it to new areas of social practice, such as protected areas for recreation and transport planning, but also by turning the focus on the human itself.

Holm reflects on the potentials of LA21 acknowledging its limitations. In spite of technocratic normalisation and declining participation in parts of the LA21 landscape, a culture and political infrastructure for continuous niche experimentation is well established. The most successful of these experiments have proven to be socially feasible, technically efficient and politically progressive, and capable of advancing positive feedbacks to the institutional regimes of decision making. The challenge is to advance these deliberative niche efforts to achieve more sustainable socio-technical regimes.

In the case discussed by Tolnov Clausen, Hansen and Tind, the conflict between strategic interests resulted in a weakness to build up democratic plans and perspectives for basic nature preservation in a local area. Authorities avoid a mobilisation of active citizens in order to keep open the doors for strategic compromises between strong stakeholders. Referring to bottom-up experiments in the local community the authors document a possible alternative political approach (which was not chosen): The authorities could stimulate a democratic process letting citizens' active dialogues define plans and perspectives for nature preservation and sustainable development in the local area. By empowering citizens' own responsibility for nature preservation the authorities could – instead of mediating between strong stakeholders – stimulate creation of community values and commons. Open democratic values does not come out of stakeholder negotiations and as long as sustainability is defined in a language of conflict between strategic interests it does not contribute to shared values for community development.

Bransholm Pedersen and Land argue in their chapter how linking the concepts of sustainability and health can expand the current more limited focus of health in terms of health care services. Such linking can contribute to conceptualising health as part of a normative horizon of orientation for better condition of sustainability and health for the entire population. Within such a realm of understanding, they argue, societal responsibility is related to practice and experience of everyday life, which means that the dimension of sustainability can be incorporated in an everyday life practice. Health problems are lifted up to a common social responsibility and not just a problem and responsibility for the individual.

Figueroa concludes that political and planning transport decisions ought to be reconsidered in the light of basic democratic principles to facilitate a transition toward sustainable transportation in large urban areas in the Global South. Transport sustainability demands great experimentation for which democratic mobility principles can provide a fruitful platform toward formulation of more inclusive and viable mobility templates. A full reconciliation with a modern technologically driven transport development paradigm cannot be expected for the majority of the world's poorest urban population.

Emerging Directions for Sustainability Research and Milestones for Political Debate

Several lines of research concerning sustainability have emerged as a result of the efforts and discussions leading to this book. Clearly, the diversity of the contributions themselves can be taken as indication of the profile of future research. The analyses in this book illustrate a variety of attempts to renewing and reinvigorating the concept of sustainability. All in all, a new agenda for sustainability needs to give attention to a new set of defining elements among them; a holistic notion that embraces ethical considerations, sustainable rhythms of life, a new global ethics to negotiate sustainable lifestyles based on an ethics of social responsibility, a focus on human capabilities, a new valuation of our commons and how we can make sustainable use of them in a global perspective. On the practical level, the authors have asked how we can transition into sustainability from local 'niche' experiments to the society and the globe at large. The attempted answers consider ideas from within sectors: linking health and sustainability, linking democracy and sustainable transportation, food democracy, to approaches within and between diverse disciplines: social framing, policy and decision making, economic and environmental modelling.

Comparing the broad approach of our results with a so-called strong concept on sustainability advanced by biologists (clear limits for the reproduction of nature within societal practices) our contributions point to a much more complicated pattern for future development. Our concept of development includes contradictory trends in nature as well as culture and society and without clearly identifiable models for sustainable actions. The knowledge for future actions must come from the analysis of natural systems as well as social lifestyles and societal systems.

Compared to a concept of sustainability advanced by economists – economic growth within acceptable impacts on nature – our contributions are much more aware of the existing dialectics in-between nature and culture and that restraining societal practices in the end must be based on social criteria and not only on natural ones. The reasons behind this assertion are many, but first and foremost comes the fact that in the end, social practice and criteria for action can be predicted and directed by political systems, and it does not make sense to speak of natural developments 'in themselves'.

In what follows we elaborate a bit further on the aspects of health, interdisciplinarity, social responsibility and democracy which we consider are our stronger contributions to a new agenda for sustainability. In addition, we hope that some issues may become milestones in the political debate in the field; and we strongly believe that the significance of issues such as environmental and social responsibility should not diminish with time.

Relevance of Research into Sustainability and Health

Since the needs of the present as well as future generations are key components of the Brundtland definition of sustainability, it seems to be obvious that health must be an important aspect of sustainable development. Nobody can dispute that health is a basic human need. Health care was mentioned by the Brundtland Commission as one of the needs that must be met for development to be sustainable (World Commission on Environment and Development 1987: 55). Despite this, health has often played a secondary role in discourses on human needs and sustainability. There may be several reasons for this, both institutional and others, but in recent years there has been an increasing focus on the links between sustainability and health as evidenced for instance in a number of publications from the World Health Organization (see MacArthur and Bonnefoy 1998).

This recent stronger focus on health raises important questions about the conceptual understanding of health in relation to sustainability. How exactly should we understand health as a human need, and which needs are basic needs? More fundamentally, how do we conceptualise health and what will the implications of such a conceptualisation be for the way we see the links between health and sustainability?

In this book, Crabtree proposes a list of essential capabilities, rather than needs, which in relation to the individual include bodily, mental and emotional health. These are necessary for individuals to lead the lives they value. His approach emphasises the importance of mental health and its relationship with the environment and why this is a key aspect of sustainability. Bransholm Pedersen and Land suggest that health and sustainability should be understood as a duality in order to capture their mutual impact and the 'demands' they put on each other. They base their discussion on the re-conceptualisations of both the concept of health and of sustainability that have been proposed in recent years. Finally, Hvid's basic point is that sustainable rhythms in modern industrial society have been detached from each other and have been reduced to either 'controlled repetition' or 'discontinuous chaos'. This has profound implications for human capacities and for both physical and mental health. The big challenge for sustainability, in Hvid's conception, is to reunite the spheres of repetition and discontinuity into a coherent everyday life of sustainable rhythms.

In summary, these contributions advance a strong argument for why the issue of health needs to become an example of social responsibility. Linking health to sustainability is necessary because modern health is not pending solely on individual matters, but is intrinsically connected to how people exercise their social responsibility and by leading forms of life that makes health and sustainability achievable.

Relevance of Interdisciplinary Research Towards Sustainability

The relevance of interdisciplinary research is well illustrated by the contributions presented in this book. A few examples from the contributions exemplify the importance of this interdisciplinary approach in research for sustainability. Figueroa calls attention to exploring new paradigmatic horizons for transport development in the South departing from considerations of democratic principles that can open up the spectrum of solutions and the possibility for the so-called informal solutions to transport: 'street vendors, rickshaws, non-motorised vehicles, to find a place for becoming a viable part of sustainable mobility in the Global South'. In another example, Brandt's analysis shows how the old land use system did not ensure sustainability, but quite the opposite, an official blindness to overexploitation, and that the sea fishery was not influenced by the culturally accepted land use system or the international science-based valuations of fish stocks and optimal fish quotas.

Both of these very different examples addressing the complexity of reality demonstrate the intimate way in which natural, social, technological, cultural, political and economic factors are connected, and that strategies for sustainable development must be based on a thorough understanding of the complexity of political, social and natural conditions and their mutual contingencies.

Relevance of Research Emphasising Social Responsibility

Another direction for the type of research and political milestone that this book seeks to exemplify emphasises the role of *social responsibility* as opposed to the role of market-oriented solutions in achieving sustainability. As previously mentioned, social responsibility but mostly at a corporate level as in corporate social responsibility (CSR) has become a prominent issue. Despite examples of the use of CSR as a platform for branding or downright window dressing embraced by businesses and governments alike, the idea should be that such solutions create public space for the emergence of a wider socially responsible approach to sustainability. However, since CSR is by definition necessarily limited by market constraints it remains to be demonstrated the extent to which it is possible to expect that corporate responsibility acts as a precursor to furthering social responsibility.

A different approach is presented by Elling who considers the question of establishing social responsibility connected to the differentiation between spheres of rationality: the cognitive-instrumental, the ethical and the aesthetic rationalities – that have been outdifferentiated in the process of modernisation. Social responsibility is for Elling's argumentation a question of public mediation between out-differentiated spheres of rationalities, therefore he emphasises the need for establishing public arenas for communication.

Summing up, there is a need to further advance critical research into the question of social responsibility. Under what conditions can social responsibility be developed in various contexts? What happens when 'the social' is reduced to costs and benefits in economic terms? And, how are the social needs and livelihoods of people affected within the framework of 'development' understood as economic growth? And, vice versa, how can social needs and livelihoods of people affect economic developments and trendsetting?

Another implication of this line of research will lead us to critically re-examining the potentials as well as the contradictions inherent in the demand for changing the quality of economic growth, as the Brundtland Commission formulated it. This could have implications for the arrangements taking place regarding international trade in CO_2 quotas within the framework of the Kyoto agreements. As for today, mostly market-oriented solutions are favoured in the numerous political programmes for mitigation of the climate problems at national as well as international level. The most recent events regarding climate, society and global markets has challenged this paradigm. The environmental effects and the social implications of such arrangements are today principally linked to market mechanisms and they need to be also systematically addressed in terms of social responsibility. These tasks can also be framed in terms of the discussion about: 're-inventing the commons'. This should not be understood as a return to tradition but rather, since local contexts today are inevitably connected with the global, as the enabling, recognition and display of common goods at all levels from the local to the global (Nielsen and Nielsen 2006: 309, Tolnov Clausen, Hansen and Tind in this volume).

Relevance of Research Linking Sustainability and Democracy

The last area of research emphasis considers *democracy and social dialogue* as a prerequisite for sustainable development. This issue is dealt with in a number of the contributions throughout the book; for example: Hvid, Elling, Hansen, Søndergård and Stærdahl, Jelsøe and Kjærgård, Tolnov Clausen, Hansen and Tind, Holm, Figueroa and Crabtree. Tolnov Clausen, Hansen and Tind insist that sustainability must be a democratic project (cf. above), and the results of their research demonstrate the relevance of wider participation towards sustainability, beyond or despite all the difficulties arising from the clash between rhetorical emphasis on democracy and citizen participation and technical rationality that underlie many of the actions and attitudes of the institutional and political systems. Similarly, Figueroa emphasises the use of the term 'practices of democratic politics' as a civic culture of equality, association, participation and mobilisation that are better able to illuminate sustainability thinking and practices rather than a discussion of democracy and sustainability merely connected to a process of decision making.

Holm shows that experiments with LA21 in Denmark have been associated with efforts to green institutions, such as everyday life experiments on new partnerships, citizens' participation, practical experiences in car-sharing, green purchasing, public provision of organic meals, etc. These experiments have enhanced new identities, cognitive frameworks and practical knowledge. However, he also warns that current trends towards technocratic normalisation and declining participatory efforts constitute a threat to the deliberative part of the political modernisation in LA21. Holm, as well as Hansen, Søndergård and Stærdahl introduce the concept of transition arena, referring to a multi-actor network formed in relation to specific sectors or innovation systems with the aim of establishing new rationalities and capacities. Hansen, Søndergård and Stærdahl also discuss some of the problems associated with power and the top-down approach of incumbent actors within socio-technical systems and how to make more room for different and conflicting interpretations of what sustainable futures could look like. The advanced management of innovation processes alone does not facilitate a genuine transition towards sustainable practices. Such a transition requires taking politics very seriously, and this inevitably enhances the conflictual nature of the transition management.

A last example, taken from Jelsøe and Kjærgård considers, with reference to Lang and Heasman (2004), the concept of food democracy, as opposed to food control, and argue for the need of involving the population more widely in the development of food production and control. Some of the barriers of change towards sustainable development of the food sector are intimately connected to the dominant trends in current global development: for example the dominance of large supermarket chains with their capacity to 'internalise consumer demand' (Wilkinson 2007). A change of this deadlock would require involvement of the population as 'consumer-citizens' who seek to have a more direct influence on the development of new practices and initiatives as well as to create pressure for new forms of social regulation within the food sector.

All these examples point towards the need for continued research activities that can evaluate experiences with democratic participatory activities for the promotion of sustainability and facilitate new and permanent experimentation as well as discussion of new visions for the future.

The formulation of a new agenda for sustainability cannot be accomplished with a simple move or seeking a new principle that can replace an old agenda. In fact almost all dimensions that need to be part of a future agenda are already present in research and current international debates but lack practical implementation. At least one new issue – the climate issue – seems to actualise an intensified need to create a connectedness between several dimensions of sustainability. As argued by Söderbaum (2008) such connectedness does not any longer come out of a meeting with neoclassic economic theory and in the book we propose other connecting discourses like democratic values, ethics and institutional regulation.

In relation to climate our shared destiny has become even more evident than in earlier sub-issues in sustainability research and politics. The climate change challenges demand a synchronous attention to universal and contextual questions. In politics as well as in research, ethical or normative issues are necessary to make the connectedness of different sustainability sub-questions possible; and the universal ethical and normative questions become abstract and empty in today's complex global world without a specific research in contexts and in relating to specific social, cultural and individual realities. Complexity encloses many conflicts and incalculable uncertainties which need as well to be met by decentralised citizens' responsibilities as global legitimate regulation. With this book we have argued that a future agenda for sustainability needs to critically walk that balance.

References

Lang, T. and Heasman, M. 2004. *Food Wars: The Global Battle for Mouths, Minds and Markets.* London: Earthscan.

MacArthur, I. and Bonnefoy, X. 1998. *Environmental Health Services in Europe 2: Policy Options.* WHO Regional Publications, European Series, No. 78. Copenhagen: World Health Organization, Regional Office for Europe.

Nielsen, K.A. and Nielsen, B.S. 2006. *En Menneskelig Nature.* Frederiksberg: Frydenlund.

Søderbaum, P. 2008. *Understanding Sustainability Economics.* London: Earthscan.

Wilkinson, J. 2006. Fish: A Global Value Chain Driven onto the Rocks. *Sociologia Ruralis*, 46(2), 139–53.

World Commission on Environment and Development. 1987. *Our Common Future.* Oxford: Oxford University Press.

Index